Physical Science in the Middle Ages

HISTORY OF SCIENCE

Editors

GEORGE BASALLA
University of Delaware

WILLIAM COLEMAN
Johns Hopkins University

Biology in the Nineteenth Century:
Problems of Form, Function, and Transformation

WILLIAM COLEMAN

Physical Science in the Middle Ages

EDWARD GRANT

The Construction of Modern Science:
Mechanisms and Mechanics

RICHARD S. WESTFALL

Physical Science in the Middle Ages

EDWARD GRANT

Department of History and Philosophy of Science
Indiana University

CAMBRIDGE UNIVERSITY PRESS

CAMBRIDGE

LONDON · NEW YORK · MELBOURNE

Published by the Syndics of the Cambridge University Press
The Pitt Building, Trumpington Street, Cambridge CB2 1RP
Bentley House, 200 Euston Road, London NW1 2DB
32 East 57th Street, New York, NY 10022, USA
296 Beaconsfield Parade, Middle Park, Melbourne 3206, Australia

Library of Congress catalogue card number: 77-083993

ISBN 0 521 21862 4 hard covers
ISBN 0 521 29294 8 paperback

First published by John Wiley & Sons, Inc., 1971
First published by Cambridge University Press, 1977

Printed in the United States of America
Printed and bound by Halliday Lithograph Corp.,
West Hanover, Mass.

To Robyn, Marshall, and Jonathan

Series Preface

THE SCIENCES CLAIM an increasingly large share of the intellectual effort of the Western world. Whether pursued for their own sake, in conjunction with religious or philosophical ambitions, or in hopes of technological innovation and new bases for economic enterprise, the sciences have created distinctive conceptual principles, articulated standards for professional training and practice, and brought into being characteristic social organization and institutions for research. The history of the sciences—astronomy; physics and associated mathematical methods; chemistry; geology; biology and various aspects of medicine and the study of man—consequently exhibits both great interest and exceptional complexity and presents numerous difficulties for investigation and interpretation.

For over half a century an international group of scholars have been studying the historical development of the sciences. Such studies have often called for an advanced level of scientific competence on the part of the reader. Furthermore, these scholars have tended to write for a small specialist audience within the history of science. Thus it is paradoxical that the ideas of men who are professionally committed to elucidating the conceptual development and social impact of science are not readily available to the modern educated man who is concerned about science and technology and their place in his life and culture.

The editors and authors of the *History of Science Series* are all dedicated to bringing the history of science to a wider audience. The books comprising the series are written by men who are fully familiar with the scholarly literature of their subject. Their task, and it is not an easy one, is to synthesize the discoveries and conclusions of recent scholarship in history of science and present the general reader with an accurate, short narrative and analysis of the scientific activity of major periods in Western history. While each volume is complete in itself, the several volumes taken together will offer a comprehensive general view of the Western scientific tradition. Each volume, furthermore, includes an extensive critical bibliography of materials pertaining to its topics.

George Basalla
William Coleman

Preface

IN ITS MOST MEANINGFUL SENSE, the history of medieval science is the history of the dissemination, assimilation, and reaction to ancient Greek science as it passed from the Byzantine Empire to Islam and subsequently to Western Europe. Although I am fully cognizant of the enormous scientific debt owed by medieval Europe to Arab and, to a lesser extent, Byzantine civilization, my aim here is to describe briefly the significant scientific developments and interpretations formulated in Western Europe from the period of the Late Roman Empire to approximately 1500 A.D. To describe these developments comprehensively over the whole range of the particular sciences would require a volume—more likely volumes—of far greater scope than envisioned here. Fortunately, the content and concepts that came to dominate medieval science from the late twelfth century on were powerfully shaped and dominated by the science and philosophy of Aristotle (384–322 B.C.). His explanations and interpretations of the structure of the world and its physical operations were so pervasive that an understanding and grasp of the range of problems and solutions to which his writings gave rise would provide the reader with a genuine insight into the nature, achievements, and failures of medieval science. For this compelling reason, after characterizing early medieval science in the first two chapters, I have focused attention on problems and controversies associated with Aristotelian physical science in the Later Middle Ages.

Aristotle, or members of his school, had something to say, either superficially or in depth, on virtually all the sciences studied in the Middle Ages. A large body of technical and specialized scientific literature was also elaborated in late antiquity after the time of Aristotle but many centuries before it reached Western Europe. A detailed description of these developments would be beyond the scope of this volume. However, where appropriate and relevant, I have included essential features and ideas from this large body of scientific literature that were either incorporated into medieval Aristotelian science or rejected in the controversies that raged incessantly in the medieval universities. Hopefully, this procedure will also allow some larger sense of the history of medieval science to emerge. My primary objective, however, is to convey a proper sense of the general impact of Aristotelian thought and the medieval reaction to a few of the major physical and cosmological problems that emerged from the works of Aristotle. The medieval preoccupation with Aristotle's physics and cosmology and their centrality in shaping the medieval world view amply justify this approach.

Edward Grant

Contents

Physical Science in the Middle Ages

CHAPTER I

The State of Science from 500 A.D. to 1000 A.D.

FROM THE TIME when Greek philosophy and science first penetrated the Roman world during the second and first centuries B.C. to the present, it is an indisputable fact that science reached its lowest ebb in Western Europe between approximately 500 A.D. and 1000 A.D., improving gradually until the influx of Greek and Arabic scientific treatises in the 12th and early 13th centuries introduced a virtually new body of scientific literature. How did such a disastrous state of affairs arise and what perpetuated it for so many centuries?

Because the period in question was preceded by the gradual disintegration and transformation of the Roman Empire and the triumph of Christianity as a state religion, these events almost inevitably serve as the large historical background against which the decline of science must be viewed. As early as the reign of Diocletian (285–305), the political instability of a few centuries had seen the Roman Empire divided into Eastern and Western halves, a division that became irreparable after the death of Theodosius in 395 A.D. During the course of the fifth century, the Western half fell prey to invading Germanic tribes and by 500 A.D. much of it was in their control. Despite subsequent efforts of the Eastern emperor, Justinian, only the trappings of Empire remained—the substance was dead, and Western Europe evolved new forms of social and governmental activity to cope with conditions drastically different from those of a few centuries earlier. With a breakdown of strong central government and the gradual dissolution of the urban life so characteristic of the first few centuries of the Empire, it is hardly surprising that intellectual life in the Western

1

half suffered. If a reasonable degree of political stability, urban activity, and patronage of some kind, are essential or at least conducive to the pursuit of science, the absence of these enables us to comprehend, in a quite general way, how scientific understanding and achievement could have deteriorated and stagnated over so long a period of West European history.

The triumph of Christianity was, among other things, the culmination of a struggle and competition between mystery religions and cults that began as far back as the Hellenistic period and continued until the emperor Theodosius declared Christianity the only legal religion in 392 A.D. As economic and political oppression grew increasingly burdensome to great masses of people at all levels of society, the mystery religions gained in favor and their doctrines were easily disseminated via the excellent roads linking the far-flung points of the Roman Empire. The cults of Isis, Mithra, Cybele, Sol Invictus (Unconquered Sun), as well as Gnostics, Christians, and others, not only borrowed ideas and rituals from one another but also came to share a few basic beliefs. The world was evil and would eventually pass away. Man, sinful by nature, could achieve immortal bliss if only he turned away from the things of this world and cultivated those of the eternal spiritual realm. Along with varying degrees of asceticism, many of the cults believed in a redeemer god who would die in order to bring eternal life after death to his faithful followers. Even some of the contemporary philosophic schools, such as Neo-Platonism and Neo-Pythagoreanism, sought to guide their adherents toward salvation and union with God and, although they utilized more intellectual means, they were not above employing magic to achieve their ends.

Indeed, acceptance of magic and occult powers was widespread in the Roman Empire during the first few centuries of the Christian era, as evidenced by numerous treatises ascribed to the Egyptian god, Thoth, known to the Greeks as Hermes Trismegistus ("Thrice-great Hermes"). Although it incorporated elements from a variety of current philosophies such as Platonism, Neoplatonism, Stoicism, and others, and drew upon some aspects of contemporary scientific knowledge and theory, Hermetic literature represented a reaction to the traditional rational approach of Greek philosophy and science, for it sought to comprehend and explain the universe by magic, intuition, and mysticism. Because these treatises were attributed to the god Hermes and emphasized Egyptian wisdom, readers uncritically accepted them as works of great antiquity, antedating Plato and perhaps even Moses. They were the repositories of a pristine source of ancient wisdom and as such exerted an enormous influence. Even Church Fathers read and admired them. Lactantius (*fl.* 300 A.D.), who read them in the original Greek, was highly respectful toward Hermes, whom he regarded as a Gentile prophet of Christianity. St. Augustine, who read at least one of the treatises in Latin translation and rejected a description of the animation of statues of Egyptian gods by magical means, fully accepted Hermes as one who had exerted a strong moral force in Egypt after the time of Moses but long before the ancient philosophers and sages of Greece. Although a few of the Hermetic treatises were available in Latin translation and exerted an influence during the Middle Ages, their full impact would come in the Renaissance, when they provided a widely accepted guide to the study and appreciation of nature and religion.

Did this intense and widespread search for other-worldly salvation, in which the physical world was either held in contempt or approached by means of magic and occult forces, engage the minds and energies of those who, in an earlier age, would have devoted their talents to science and mathematics? If so, it is not easily detectable, at least not before the victory of Christianity. Indeed, during the first few centuries of the Roman Empire, when Christianity was relatively weak and uninfluential, struggling for survival against its many rivals, some of the greatest scientific works of the ancient world were written (as always in the Greek language), a few of which would exert profound influence on the later course of medieval science and well beyond into the Renaissance.

The first century A.D. saw the significant works of Hero of Alexandria (who wrote on pneumatics, mechanics, optics, and mathematics), Nicomachus (on Pythagorean arithmetic), Theodosius and Menelaus (who wrote on spherical geometry; Menelaus' *Spherics* is especially important for the treatment of spherical triangles and trigonometry). The heights were reached in the second century when Claudius Ptolemy wrote the *Almagest,* the greatest treatise in the history of astronomy until the time of Copernicus in the sixteenth century, as well as technical works in optics, geography, stereographic projection, and even the greatest of all astrological works, the *Tetrabiblos* (known in Latin as the *Quadripartitum,* the four-parted work). In the medical and biological sciences, Galen of Pergamum produced about 150 works embracing both theory and practice. His works formed the foundation of medical theory and study until the sixteenth and seventeenth centuries. Even in the third century significant contributions were made in mathematics by Diophantus in algebra and later by Pappus, who not only wrote commentaries on the great mathematical works of Greek antiquity but, in his *Mathematical Collection,* showed originality and understanding of a high order. These achievements, spread across three to four centuries, were typical of the manner in which Greek science had developed and advanced. Always the product of a small number of men concentrated in a few centers, Greek science was a fragile enterprise able to advance and preserve itself just so long as the intellectual environment was favorable, or at least not overtly antagonistic.

With the triumph of Christianity in the fourth century, that small but essential handful of men, who in previous centuries had somehow managed to comprehend, advance, and perpetuate an inherited body of high-level theoretical science, was no longer produced in the eastern or western parts of the Empire (because Greek was the language of the eastern half and some of the scientific treatises could be read in the original language, a much higher level of comprehension was maintained there; but the spark of originality had flickered out). By 500 A.D., the Christian Church had drawn most of the talented men of the age into its service, in either missionary, organizational, doctrinal, or purely contemplative activity. Honor and glory were no longer found in objective, scientific comprehension of natural phenomena, but rather in furthering the aims of the universal Church.

The intense and bitter polemic against pagan learning and religion, which had marked the long struggle of Christianity, cast a pall of suspicion over Greek philosophy and science. In its moment of triumph, Christianity looked with fear and distrust, if not with downright hostility, upon its fallen foe. But Christians were hardly of a single mind on the matter. The most extreme reaction was represented by Tertullian (*ca.* 160–*ca.* 240) who viewed philosophers as purveyors of damnation and heresy. Any alliance between Athens and Jerusalem was unthinkable. Perhaps more truly representative were those like Justin Martyr (*d. ca.* 163–167) and Clement of Alexandria (*ca.* 150 and *d.* before 215) who looked upon Greek learning and philosophy as the handmaid of theology, to be used for a better understanding of the Christian religion, but not to be studied for its own sake. For just as philosophy had prepared the Greeks to accept Christianity and the perfection of Christ, so might it perform the same good work for others. The Christian dilemma is best illustrated by St. Augustine, whose influence throughout the Middle Ages was enormous. In 386 A.D. he emphasized the importance of the liberal arts which from the time of

classical Greece had included the four sciences of geometry, arithmetic, astronomy, and music. These traditional disciplines were helpful in leading a good life and indispensable for a proper understanding of the universe. Augustine even contemplated the composition of an encyclopedia of the liberal arts which would have included sections on the scientific disciplines mentioned above. Only a small part of this project was ever written, perhaps because in later life his attitude toward pagan and secular learning underwent drastic alteration. A few years before his death, he bitterly regretted his earlier emphasis on the liberal arts and concluded that the theoretical sciences and mechanical arts were in no way useful to a Christian.

Despite obvious concern and trepidation about the potential dangers of pagan learning, of which science and philosophy were integral parts, circumstances forced an uneasy truce and compromise. Virtually the only secular learning available was of pagan origin. Elementary and advanced instruction were permeated with pagan religious, philosophic, mythologic, and literary references. Illustrations in grammar and rhetoric texts were drawn wholesale from pagan sources. Christians who received formal secular instruction inevitably absorbed large quantities of traditional pagan fare. Reluctantly, the Church found it necessary to modify its attitude, if not its uneasiness, about pagan learning and science. Indeed, the account of creation in *Genesis* required Christians to provide a measure of physical explanation about the world, as evidenced by the numerous commentaries on the six days of creation, or hexameral treatises, as they were called, which began to appear in the fourth century. By the fifth and sixth centuries some Christians began to manifest a degree of interest in science, and we must now inquire about the scientific treatises and texts that were available and from which their knowledge and opinions about the world were drawn.

It will be evident in what follows that the societal forces operating to weaken and dilute interest in science in late antiquity were aided and abetted by an independent process whose modest beginnings are clearly discernible as far back as the Hellenistic age (320–30 B.C.) and which would continue unabated through the first five or six centuries of the Christian era. I refer to the handbook and encyclopedic tradition of learning whose objective it was to popularize and disseminate the theories and results—but not the technical content or procedures—of Greek science.

The glorious scientific achievements of classical Greece, climaxed by the monumental contributions of Aristotle, were deepened and ad-

vanced during the Hellenistic age when the works of Euclid, Archimedes, Appollonius of Perga, Hipparchus, and others, in the physical sciences were matched by equally significant contributions in medicine and the biological sciences by the likes of Theophrastus, Herophilus, and Erasistratus. As we have already seen, work at this level continued to the fourth century A.D. But, as in our own day, there must have existed a large educated audience keenly interested in the physical world but with little inclination or ability to tackle the forbidding theoretical works at the highest level. To meet the needs of this group, a host of scientific popularizers predigested and made palatable the technical results of the various sciences which were then incorporated into handbooks and manuals. It comes as no surprise to learn that some of these treatises were filled with conflicting and contradictory reports leaving the reader to reconcile them as best he could.

Greeks who were instrumental in shaping the handbook tradition were the polymath Eratosthenes of Cyrene (*ca.* 275–194 B.C.), who supplied much geographical knowledge to the tradition, Crates of Mallos (*fl.* 160 B.C.), and especially Posidonius (*ca.* 135–51 B.C.), whose numerous works have not survived, but whose opinions on meteorology, geography, astronomy, and other sciences were absorbed into later handbooks to become permanent fixtures in the tradition. Continuing in the manner of Posidonius were other Greek authors such as Geminus (*ca.* 70 B.C.), Cleomedes (first or second century A.D.), who wrote an astronomical and cosmological work *On the Cyclic Motions of the Celestial Bodies,* and Theon of Smyrna (first half of the second century A.D.), who wrote a *Manual of Mathematical Knowledge Useful for an Understanding of Plato* in which the whole universe is discussed, as in Plato's *Timaeus,* drawing information from Hellenistic astronomy, cosmology, and Pythagorean arithmetic and mathematics. Commentaries on Plato's *Timaeus* constituted a significant part of the handbook tradition from the Hellenistic period to the early Middle Ages. Since the *Timaeus* was a scientific treatise concerned not only with the cosmos but also with the physical structure and functions of man, it was an admirable vehicle for a handbook discussion since physical and biological material could be appropriately included.

When, as a consequence of their conquest of Greece, Roman gentlemen were brought into contact with Greek culture during the course of the second and first centuries B.C., the Greek handbook tradition was firmly established and its treatises were admirably adapted to cater to Roman cultural interests. For although the Romans were impressed and

awed by Greek intellectual accomplishments, they had no interest in theoretical and abstract science. Hence when fashion dictated that cultured Romans acquire a nodding acquaintance with the results of Greek science, the handbook method was readily available. Undoubtedly some Romans, who learned Greek, could consult the Greek handbooks directly, but the great majority probably absorbed their knowledge through Latin translations. Soon, Romans themselves began compiling their own handbooks on science, and, not surprisingly, these were inferior to their Greek counterparts.

Although the Latin encyclopedic tradition actually began in the first century B.C. with Marcus Terrentius Varro (116–27 B.C.), its two most significant early representatives were Seneca (*d.* 68 A.D.) and Pliny the Elder (23/24–79 A.D.). In his *Natural Questions,* Seneca concerned himself largely with geography and meteorological phenomena (for example, rainbows, halos, meteors, thunder, and lightning) after the manner of Aristotle's *Meteorology.* He drew heavily upon Aristotle, Posidonius, perhaps his major authority, Theophrastus, and other Greek sources. Since Seneca frequently drew morals from natural phenomena, his book was popular with Christians. Of importance was the fact that it passed into the Middle Ages an estimate of the size of the earth that was small enough to encourage men like Columbus and others to think that the oceans were sufficiently narrow to be readily navigable. In it was also struck an optimistic note on the progress of science and knowledge when Seneca predicted that continuous research would reveal nature's secrets.

Pliny's *Natural History* in 37 books was a remarkable scissors and paste collection of enormous scope and detail. By his own estimate, he examined about 2000 volumes drawn from 100 authors. In Book I, Pliny presents a detailed outline of the topics and a full list of the authorities used for each of the 36 volumes that follow. Thus did he honor, rather than plagiarize, his predecessors. A total of 473 authors are listed of whom presumably the 100 mentioned above were primary and some of the others known either through intermediaries or perhaps used cursorily for isolated bits of factual information. Book II is devoted to cosmography, Books III to VI to regional geography, Book VII to human generation, life, and death, Books VIII to XXXII are concerned with zoology and botany, including fabulous animals and the curative powers associated with animals and plants, and Books XXXIII to XXXVII consider mineralogy.

As an indefatigable compiler, Pliny emphasized the curious and the odd in natural phenomena. Although confusions, inconsistencies, and misunderstandings abound throughout, the weakest sections are those which involve attempted explanations of Greek theoretical science, which Pliny scarcely comprehended.

If Pliny's work was confused and frequently inconsistent, it was at least the product of great diligence coupled with an honest respect for the sources which provided the grist for his insatiable mill. With a few notable exceptions, his successors shared little of his finer instincts. In their compilations plagiarism and incomprehension became characteristic features. Thus Solinus, who lived in the third or fourth century A.D., compiled an encyclopedic work titled *Collection of Remarkable Facts,* the most remarkable fact being that it was largely plagiarized from Pliny. Solinus, in turn, was so thoroughly plagiarized that modern scholars are frequently unable to determine whether Pliny or Solinus was the source of this or that later opinion. Encyclopedic authors looked upon available handbooks as store houses of information in the public domain which could be plundered, embellished, and rearranged to suit their purposes. The final products were then paraded as learned treatises drawn directly from the original sources. The scientific works and opinions of such great figures as Plato, Aristotle, Archimedes, Euclid, Theophrastus, and others, were cited repeatedly in the handbooks as if the compiler had direct knowledge of them. In almost all instances, however, it is painfully evident that he had no direct acquaintance with the great scientific authors of the past and was but repeating—and very likely distorting—what a slightly earlier compiler had already repeated and distorted from his predecessors.

Between the fourth and eighth centuries, encyclopedic authors produced a series of Latin works that were to have significant influence throughout the Middle Ages, especially prior to 1200. Among this group, the most important were Chalcidius, Macrobius, Martianus Capella, Boethius, Cassiodorus, Isidore of Seville, and Venerable Bede. Chalcidius (*fl. ca.* fourth century A.D.) translated most of Plato's *Timaeus* into Latin and added a commentary whose astronomical portions he plagiarized from Theon of Smyrna's *Manual.* Macrobius (*fl.* 400 A.D.), a Neoplatonist, incorporated encyclopedic learning in a commentary on Cicero's *Dream of Scipio,* which is actually Book VI of Cicero's *Republic.* Martianus Capella (*fl.* 410–439) wrote the popular *Marriage of Philology and Mercury,* an ornate, florid account of the seven liberal arts and a pale reflection of classical learning and wisdom. Boethius (*ca.* 480–524) was one of the best of the Latin encyclopedists and possessed a good knowledge of Greek. He wrote on the *quadrivium* (a term he may have introduced for the four mathematical sciences of the seven liberal arts) but only the treatises on music and Pythagorean arithmetic survive, the latter in the form of a free translation of Nicomachus' *Introduction to Arithmetic.* To these he added his translations of some of Aristotle's logical treatises, perhaps Euclid's *Elements,* and unspecified works of Archimedes which have not survived. His commentaries on certain of the philosophical treatises that he translated and his most famous work, *On the Consolation of Philosophy,* written in prison while awaiting execution, were very influential. Cassiodorus (*ca.* 488–575) included sections on the seven liberal arts in his *Introduction to Divine and Human Readings* and was reasonably scrupulous about citing his authorities. Isidore of Seville (*ca.* 560–636), in addition to a treatise *On the Nature of Things,* compiled a vast encyclopedia called *The Etymologies.* The first three of its twenty books were concerned with the seven liberal arts; others were devoted to medicine and zoology. Finally, Venerable Bede (*ca.* 673–735) was one of the most intelligent of the Latin encyclopedists. In addition to a conventional encyclopedia, *On the Nature of Things,* he wrote two treatises, *On the Division of Time* and *On the Reckoning of Time,* which were concerned with calendar reckoning and considered such topics as chronology, astronomy, calendrical computations, Easter tables and the tides. Although he borrowed heavily from his predecessors, especially Isidore, Bede was capable of adding intelligently to his meager inheritance. For example, he formulated the concept of "establishment of the port," and recorded that the tides recur at approximately the same time at a particular place along the coast, although the times of occurrence vary from place to place.

Taken together, these books contained virtually the sum total of general scientific fact and comprehension through the early Middle Ages. They confronted subsequent authors with an unsystematic, chaotic, and conflicting mass of frequently irreconcilable and incomprehensible information above which few could rise until new scientific learning became available from Arabic and Greek sources. As an illustration of the confusions which abounded in the scientific literature available in the early Middle Ages let us consider an astronomical problem involving the motions of the sun, Mercury, and Venus and the fixed order of the planets. By the fourth century B.C., it had already been observed that Mercury and Venus were always seen as morning or evening stars never farther from the sun than approximately 29 and 47 degrees respectively, whereas Mars, Jupiter, and Saturn were observable at any angular distance from the sun. To interpret these astronomical facts, Heraclides of Pontus (*ca.* 388–*ca.* 310 B.C.) had argued that while Mars, Jupiter, and Saturn revolved directly around the earth as their physical center, Mercury and Venus, the inferior planets, were exceptions revolving directly around the sun, which in turn revolved around the earth. (See Figs. 1 and 2.)

When in the seventeenth century Tycho Brahe extended this theory and assumed that *all the planets* circled the sun, which, in turn, revolved annually around the earth, it was proposed as a serious alternative to the Copernican heliocentric system. Heraclides' proposal was obviously of potential importance in the history of astronomy, for it represented a major disagreement with Aristotle's cosmology in which *all* planetary motions were assumed to have the earth as their physical center. Indeed it also entailed a denial of a single fixed planetary order with respect to the earth, since at times the order would be sun, Mercury, and Venus (Fig. 1) and at other times Venus, Mercury, and sun (Fig. 2).

Knowledge of the Heraclidean system has been reconstructed from four Latin authors, three of whom are the encyclopedists, Chalcidius (he alone mentions Heraclides by name), Capella, and probably Macrobius. Despite their apparent acceptance of it, all discuss the fixed order of the planets, which presupposes a commitment to an unalterable order for the sun, Mercury, and Venus. Thus Macrobius prefers Plato's arrangement (earth, moon, sun, Venus, Mercury, etc.) to Cicero's (earth, moon, Mercury, Venus, sun, etc.), remaining blissfully unaware of the incompatibility of either order with the movements of Mercury and Venus above and below the sun. Similarly, Martianus Capella, in the very paragraph in which he wholeheartedly adopted the Heraclidean system, for which he was praised by Copernicus, presents the two tradi-

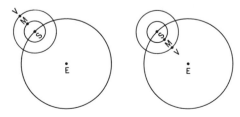

Figure 1   *Figure 2*

tional competing fixed orders of the planets. Inconsistencies of this kind could be multiplied and reveal all too clearly the frequency with which encyclopedists confounded and confused the materials which they repeated with so little understanding.

The Quadrivium or Four Mathematical Disciplines

If the level of comprehension was low, what of the *content* of science for all the many centuries of the early Middle Ages? If there was any such thing as a core of scientific learning, it would be found in the *quadrivium* of the seven liberal arts. Indeed, the four mathematical sciences (arithmetic, geometry, astronomy, and music) which comprised it were given their final condensed form by the Latin encyclopedists. Of the various accounts, the most popular and representative was compiled by Isidore of Seville in his lengthy *Etymologies*. As the title suggests, Isidore was often concerned with etymological derivations of key terms, believing that knowledge of the origin of a term conveyed an insight into the essence and structure of things.

Calling attention to the importance of arithmetic for a proper understanding of the mysteries of Holy Scripture, Isidore considers the division of numbers into even and odd, and the various subdivisions within each category. Drawing largely upon Cassiodorus, who himself had excerpted from the lengthy Boethian translation of Nicomachus' *Introduction to Arithmetic,* Isidore enunciates a melange of Pythagorean definitions including those for excessive, defective, and perfect numbers (that is, where the sum of the factors of a number exceeds, is less than, and equals, respectively, the number itself), as well as discrete, continuous, lineal, plane, circular, spherical, and cube numbers. Add to these the definitions of five types of ratio distinguished by Nicomachus, and we have virtually the whole of Isidore's arithmetic. Faced with little more than an unrelated collection of useless definitions, supplemented by a few trivial examples, the reader of Isidore's section on arithmetic could have put none of it to use. Only a comparison with the arithmetic books of Euclid's *Elements* (Books VII to IX) can illustrate the depths to which arithmetic had fallen.

Isidore has even less to say about geometry than arithmetic. Beginning with a strange four-fold division of geometry into plane figures, numerical magnitude, rational magnitude, and solid figures, he concludes with definitions of point, line, circle, cube, cone, sphere, quadrilateral, and a few others. Here we find cube defined as "a proper solid figure which is contained by length, breadth, and thickness," a definition applicable to any other solid (Euclid defines it as a "solid figure contained by six equal squares"). A quadrilateral figure is "a square in a plane which consists of four straight lines," thus equating all four-sided figures with squares!*

The longest section of the *quadrivium* is devoted to astronomy (music, like geometry, consists of a brief sequence of definitions). In a descriptive and nontechnical presentation, Isidore considers the difference between astronomy and astrology, the general structure of the universe, the sun, moon, planets, fixed stars, and comets. We learn that the sun, which is made of fire, is larger than the earth and moon; that the earth is larger than the moon; that, in addition to a daily motion, the sun has a motion of its own and sets in different places; that the moon receives its light from the sun and suffers eclipse when the earth's shadow is interposed between it and the sun; that the planets have a motion of their own; and that the stars, fixed and motionless in the heavens, are carried round by a celestial sphere, though the stars themselves are ranged at varying distances from the earth, an inference drawn from the observed unequal brightnesses of stars. Isidore believed that some of the more remote and smaller stars were actually larger than the bright stars we observe, their apparent smallness being merely a consequence of distance. It is unlikely that Isidore was even aware that an unimaginably thick and transparent sphere would be required to accommodate fixed stars varying in size and distance and distributed under the conditions described. Comprised, for the most part, of elementary and sketchy details, Isidore's astronomical discussion represents his best effort among the *quadrivium* subjects.

Isidore and his fellow encyclopedists deserve our gratitude for a valiant attempt to preserve and comprehend the tattered remnants of ancient science. But there is no denying that a scientific dark age had descended upon western Europe.

* I have slightly altered these quotations from the translation by Ernest Brehaut, *An Encyclopedist of the Dark Ages* (Columbia University Press, 1912), p. 133.

CHAPTER II

The Beginning of the Beginning and the Age of Translation, 1000 A.D. to 1200 A.D.

WITHOUT ACCESS to the hard core of Greek science, the Western world could not rise above the level of the Latin encyclopedists. While in the eighth and ninth centuries the Arabs were translating the bulk of the available Greek science into Arabic and adding to this legacy, and during the period when Greek science continued to be read and studied in the Greek-speaking Byzantine Empire, the West had before it only the rudimentary encyclopedic science already described. By 500 A.D., knowledge of Greek had become rare and knowledge of technical science even rarer. Except for occasional translations which sometimes failed to circulate or perished completely, little was added to the now dominant encyclopedic tradition. Before West Europeans would acquire serious incentive to seek out new learning from neighboring civilizations and cultures, they had first to be aroused and stimulated to a new interest in science and nature. As so often happens in the history of science, a single individual would play an instrumental role in performing this essential task.

In the late tenth century, Gerbert of Aurillac (*ca.* 946–1003), a Frenchman who became Pope Sylvester II (999–1003), used Church contacts in northern Spain to acquire a few Arabic treatises in Latin translation. From these he learned about the abacus and astrolabe, writing a treatise on the former and perhaps on the latter. The substance of his work falls squarely in the Latin tradition. Gerbert, however, was not an original thinker, and his subsequent influence was based largely on his capabilities as a teacher of science. From 972–989, he taught the seven liberal arts at the cathedral school of Reims, emphasizing a very elementary mathematics and astronomy. Laying stress on visual aids, Gerbert gained a deserved reputation as a great teacher in an age of intellectual deprivation. He not only explained how to construct a sphere to represent the heavens, but actually made one which simulated the motions of the constellations, using wires fixed on the surface of the sphere to outline the stellar configurations. Deeply impressed with Gerbert's ingenuity and dedication, his pupils went forth with great enthusiasm to continue and extend his teachings, emphasizing science as an integral part of the liberal arts. Many of the cathedral schools that rose to prominence and replaced the monastic schools as centers of learning in the eleventh and twelfth centuries were founded, or at least revived, by his pupils, the most eminent of whom were Adalberon of Laon, John of Auxerre, and especially Fulbert of Chartres. Among the cathedral schools whose beginnings or maturity are associated with his pupils are Cologne, Utrecht, Sens, Cambrai, Chartres, Laon, Auxerre, and Rouen. Until the emergence of universities in the late twelfth century, these schools were the most important centers of learning in the West.

In this school environment, intellectual interest in secular and scientific subjects was nurtured. There is ample evidence of this in an extraordinary exchange of eight letters on mathematics, sometime around 1025, between two cathedral school products, Ragimbold of Cologne and Radolf of Liège. At the initial request of Radolf, a series of mathematical questions were posed and the answers were circulated not only to the two correspondents but to others who seem to have acted as judges in what may aptly be described as a scientific tournament. Ignorant of Greek or Arabic mathematics and dependent on tidbits of geometry drawn from Roman surveying manuals or a skeletal geometric treatise ascribed to Boethius, as well as the genuine writings of Boethius, their knowledge of geometry was pitiful and fragmentary. Neither had any conception of geometric demonstration. Among other things, we find an utterly confused discussion on the meaning of exterior and interior angles in a triangle. In a question drawn from Boethius' commentary on the *Categories* of Aristotle, Radolf asked Ragimbold to calculate the side of a square which is double a given square. Although both knew that the side of the larger square is the diagonal of the smaller given square (see Fig. 3), our contestants were unaware that the sides of the two squares could not be related by such whole number ratios as 17/12 (Ragimbold's ratio) or 7/5 (Radolf's), since the two

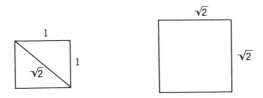

Figure 3

sides are incommensurable and therefore relatable only by an irrational ratio, $\sqrt{2/1}$ in this case. The abysmal level of understanding is of less significance than the fact that such a tournament took place in the manner described. Clearly it signifies the growing interest in scientific questions, and could hardly have occurred one hundred years earlier.

The increasing concern for intellectual pursuits brought with it a greater interest in the works of antiquity. Plato's *Timaeus,* for example, was studied intensively and used to explain the structure of the universe by the Neoplatonists at Chartres. An almost worshipful respect for ancient wisdom developed as they recognized their incalculable debt to their predecessors. If an extension of their horizons of knowledge was possible, it was only because, as Bernard of Chartres expressed it, they were privileged to stand upon the shoulders of the learned giants of antiquity, a sentiment repeated often through the centuries and even found in a letter by Isaac Newton. But the works of these giants were either unavailable or known only in a fragmentary way. As the interest in learning, especially science and philosophy, intensified during the eleventh and early twelfth centuries, the meager traditional learning was outgrown. Reports of treatises which were available in either Greek or Arabic but which were known in the West only by title, or not at all, only reinforced a sense of inadequacy. No longer willing to accept an intellectual status quo, scholars of the western world took direct action to acquire the scientific heritage of the past. The translations that followed constitute one of the true turning points in the history of western science and intellectual history in general.

Already in the middle of the tenth century, translations from Arabic to Latin were made in northern Spain at the Monastery of Santa Maria de Ripoll at the foot of the Pyrenees. These were largely concerned with geometry and astronomical instruments and were perhaps known directly by Gerbert. In the eleventh century, information about the Arab astrolabe was known to Hermann of Reichenau (1013–1054) and translations of medical treatises of Greek and Arabic authors were made from Arabic to Latin by Constantine the African, a shadowy figure associated with the medical center at Salerno in southern Italy. But the translating activity that would revolutionize western scientific thought and determine its course for centuries to come occurred in the twelfth century. Between 1125 and 1200, a veritable flood of translations rendered into Latin a significant part of Greek and Arabic science, with more to come in the thirteenth century. Not since the ninth and early tenth centuries, when much of Greek science was translated into Arabic, had anything comparable occurred in the history of science.

The great age of translation was preceded by the rollback of the Moslems in Spain and their total defeat in Sicily during the eleventh century. With the fall of Toledo in 1085 and the capture of Sicily in 1091, a now dynamic Christian Europe came into possession of great centers of Arab learning. Books in Arabic were readily at hand and intellectually starved Europeans were now eager to make their contents available in Latin, the universal language of learning in Western Europe. They came from all parts of Europe to join with native born Spaniards, whether Christian, Jew, or Arab, to engage in the grand enterprise of converting the technical science and philosophy of the Arabic language into a language that had been largely innocent of such matters. The international character of this extraordinary activity is revealed by the very names of the most significant translators such as Plato of Tivoli, Gerard of Cremona, Adelard of Bath, Robert of Chester, Hermann of Carinthia, Dominicus Gundisalvo, Peter Alfonso, Savasorda, John of Seville; and in the early thirteenth century came Alfred Sareshel (or Alfred the Englishman), Michael Scot, and Hermann the German.

Of all the Spanish centers where translations were made, Toledo became foremost. Here, and elsewhere, translations were made in a variety of ways. If the translator had mastered Arabic adequately, he would translate directly; if not, he might have teamed with an Arab or Jew. Occasionally, if he knew Spanish, he might have engaged someone to translate from Arabic to Spanish, and himself translate from the latter to Latin. A Latin translation of an original Greek treatise may occasionally have been converted through a sequence of languages, say from Greek to Syriac to Arabic to Spanish to Latin; or, perhaps, from Arabic to Hebrew to Latin. Significant distortion in the Latin end product from successive translations could hardly have been avoided.

Although the translations of the twelfth and thirteenth centuries were overwhelmingly of scientific and philosophical works—the humanities and belles-lettres were barely represented—the selection of works for translation was frequently haphazard. Availability and brevity were often decisive factors in determining whether or not a work was translated. Treatises of genuine significance were sometimes ignored while minor and occasionally trivial works were translated and subsequently studied with great intensity. Since the translators worked in widely separated places and were rarely in contact, duplication of effort was all too common. Despite great obstacles, the sum total of achievement is highly impressive. Indeed the sum total of a single translator, Gerard of Cremona (*d.* 1187), would have alone drastically altered the course of western science. As a tribute to this greatest of all western translators and as a guarantee that posterity would recognize its indebtedness and not credit to others what was Gerard's, his devoted students appended a brief biographical sketch and list of translations to Gerard's translation of Galen's *Tegni* (*Medical Art*). Here we learn that after absorbing all that was available to the Latins, Gerard went to Toledo to find Ptolemy's *Almagest,* which could not be found among the Latins. Impressed by the intellectual riches of Toledo, Gerard learned Arabic and translated not only the *Almagest,* but at least 70 more treatises. He made available the basic physical works of Aristotle (*Physics, On the Heavens and World, On Generation and Corruption, Meteorology,* Books I to III), as well as the latter's *Posterior Analytics,* the major treatise for discussion of scientific method. Among many mathematical works, he rendered Euclid's *Elements,* the *Algebra* of Al-Khwarizmi, and *The Geometry of the Three Brothers,* which included Archimedean mathematical techniques that would prove influential. In addition to many other astronomical, astrological, alchemical, and statical works, Gerard translated a large number of medical treatises, including many by Galen, the *Canon* of Avicenna, and the *Liber continens* (that is, *The Book of Divisions Containing 154 Chapters*) of Rhazes. These works alone formed the very core of medieval medical studies. To Gerard's impressive achievement could be added many other translations made by the translators mentioned above and others.

Although considerably fewer in number, significant translations were also made directly from Greek to Latin. These were done almost exclusively in Italy and Sicily where contacts with the Greek-speaking Byzantine Empire had never been broken. During the twelfth century, the Norman rulers of southern Italy and Sicily utilized these contacts to gather Greek theological, scientific, and philosophical texts. In Sicily, Plato's *Meno* and *Phaedo* were translated (by Henricus Aristippus), as was Ptolemy's *Almagest,* Euclid's *Optics, Catoptrics,* and *Data,* and a few of Aristotle's works. From the Arabic, Eugene the Emir, who was trilingual (Arabic, Greek and Latin), translated Ptolemy's *Optics.* In northern Italy, where the names of James of Venice, Burgundio of Pisa, and Moses of Bergamo, are preserved, additional translations were made from Greek to Latin. But just as Gerard of Cremona towered above all other translators from Arabic to Latin, so William of Moerbeke (*ca.* 1215–*ca.* 1286), a Flemish Dominican, was the greatest of translators from Greek to Latin. Encouraged by his friend, St. Thomas Aquinas, who had complained of the inadequacy of the translations of Aristotle's works from Arabic, Moerbeke completed new translations from Greek manuscripts of almost all of Aristotle's works except the *Prior* and *Posterior Analytics.* To these he added translations of commentaries on Aristotle's works by some of the most important Greek commentators of late antiquity such as Alexander of Aphrodisias, John Philoponus, Simplicius, and Themistius. In 1269, he translated all but a few of the numerous works of Archimedes, along with important Greek commentaries. Renaissance translators utilized these translations without acknowledgement and paid Moerbeke high tribute by publishing his translations in the first printed version of the works of Archimedes at Venice in 1503. All told, Moerbeke made approximately 49 translations ranging over theology, science, and philosophy.

Without the valiant labors of this small army of translators in the twelfth and thirteenth centuries, not only would medieval science have failed to materialize, but the scientific revolution of the seventeenth century could hardly have occurred. The mass of "new" science, so overwhelming in its scope and magnitude, had first to be absorbed, a process that occupied virtually all of the thirteenth century. Then came a period of detailed elaboration and significant alteration. By the early fifteenth century, medieval scholastic science had reached its full development founded upon an Aristotelian world view but supplemented by a wealth of specifically anti-Aristotelian criticisms made, however, within the very framework of Aristotelian science. Following a period of relative stagnation in the fifteenth and early sixteenth centuries, scholastic science would be subjected to severe criticism as significant new departures were made from it, culminating in a scientific revolution. Without the earlier translations, however, which furnished a full-blown and well articulated body of theoretical science to Western Europe, great scientific revolutionaries such as Copernicus, Galileo, Kepler, Descartes, and Newton would have had little to reflect upon and reject, little that could focus their attention on significant physical problems. Many of the burning issues and vexing scientific problems that were resolved in the seventeenth century entered Western Europe with the translations or were brought forth by medieval authors, who systematically commented upon that body of knowledge. Of this mass of science and learning, the physical and philosophical works of Aristotle were fundamental, containing as they did an overpowering and comprehensive scientific view of the cosmos that was wholly new to the West, and which was destined at times to illuminate and, at times to confuse, but at all times to dominate the minds of learned men. It is the impact of Aristotelian science and subsequent reactions to it that will form the central theme of the chapters to follow.

CHAPTER III

The Medieval University and the Impact of Aristotelian Thought

B Y 1200, THE UNIVERSITIES of Paris, Bologna, and probably Oxford, were flourishing centers of learning. Although documents that shed light on their origins and early development are virtually nonexistent until the thirteenth century, by which time they were already well established, the spontaneous emergence of universities was intimately associated with the new learning that had been translated into Latin throughout the course of the twelfth century. Indeed the university was the institutional means by which western Europe would organize, absorb, and expand the great volume of new knowledge; the instrument through which it would mold and disseminate a common intellectual heritage for generations to come. While the universities of Paris and Oxford became renowned as centers of philosophy and science and Bologna for its schools of law and medicine, all three, as well as approximately eighty universities founded subsequently and patterned after Paris (in northern Europe) and Bologna (in southern Europe), shaped the university into a form that has persisted to this day. The medieval university was an association of masters and scholars subdivided into faculties (primarily arts, law, medicine, and theology) in each of which a formal and required curriculum was pursued toward bachelor's and master's, or doctoral, degrees. Although unrecognizable to ancient Greeks, Romans, and Arabs, it would be wholly familiar to students and faculty at any of our modern universities. Residential colleges, especially at Oxford and Cambridge where they became the basic intellectual unit, were already in existence in the late twelfth century and multiplied during the thirteenth (one of these, Merton

20

College at Oxford, was destined to play a major role in the history of medieval science during the fourteenth century). By 1500 approximately sixty-eight colleges had been established in a variety of universities.

The gradual introduction of the new learning rendered obsolete the meager traditional curriculum of the cathedral schools, a curriculum that had preserved a balance between science, literature, and the humanities. As the universities wholeheartedly embraced the new philosophic and scientific knowledge, they forged a new and greatly expanded curriculum which destroyed this balance. By the mid-thirteenth century, the arts courses required for the degree of Master of Arts, a degree that was prerequisite for study in the higher faculties of law, medicine, and theology, were heavily oriented toward logic and natural science. The logical, scientific, and philosophical works of Aristotle now formed the hard core of the curriculum. The required program of study culminating in the Master of Arts degree at the universities of Paris and Oxford was not, as those unfamiliar with the Middle Ages may suppose, top-heavy with courses in theology and metaphysics. Rather, it consisted for the most part of courses in logic (which had absorbed much that was grammatical), physics, which embraced physical change of all kinds, cosmology, and elements of astronomy and mathematics. Since virtually *all* students in arts studied a common curriculum, it becomes clear that higher education in the Middle Ages was essentially a program in logic and science. Never before, and not since, have logic and science formed the basis of higher education for all arts students.

Leaving aside Aristotle's logical works, which were carefully studied, the following scientific treatises by him were fundamental: *Physics* (in eight books), which was devoted to the conditions and principles of change as well as motion in general; *On the Heavens and World* (*De caelo et mundo;* in four books), which was concerned with the motions of heavenly and terrestrial bodies; the *Meteorology* (in four books), which described and explained a wide range of phenomena thought by Aristotle to occur in the uppermost reaches of the terrestrial region just below the sphere of the moon (for example, wind, rain, thunder, lightning, and even comets and the milky way, which Aristotle judged to be meteorological rather than astronomical phenomena); and *On Generation and Corruption,* which considered transmutation of the four basic material elements, as well as chemical change involving elements and compounds. Also in the curriculum were Aristotelian treatises on biology, metaphysics, psychology, and ethics. In astronomy, the most popular works were the anonymous *Theory of the Planets* (*Theorica planetarum*), written around 1200 and containing a series of definitions of technical terms, and John of Sacrobosco's *On the Sphere* (*De sphera*). In mathematics, Euclid's *Elements,* Books I to VI, were especially required for geometry, while Boethius' *Arithmetic* and Euclid's *Elements,* Books VII to IX, served as texts for the study of theoretical arithmetic, or number theory. Familiarity with arithmetic operations (addition, multiplication, subtraction, and division) was conveyed by means of applied, or practical, arithmetics, among which the most popular was the *Algorism* of John of Sacrobosco. Although some of the basic texts used in courses in astronomy and mathematics are known from extant curriculum lists, the manner of teaching these subjects in the classroom and the actual content of the courses are, at present, virtually unknown.

Most of the scientific concepts, theories, and disputes that will be mentioned or discussed below, are contained in commentaries on these and other standard texts required for study in the universities. One form of commentary involved the systematic exposition of a text— frequently some work by Aristotle—in which a portion of the text was presented, followed immediately by the commentator's explanation of its meaning with occasional interjections of his own opinions or interpretations. Thomas Aquinas employed this form in most of his Aristotelian commentaries.

But the more customary, and far more significant, method of considering the content of the standard school texts was in the form of questions or problems, *Questiones* as they were called. Although the presentation of scientific problems in question form can be traced back to antiquity (for example, the *Problems,* falsely attributed to Aristotle, and the *Natural Questions* of Seneca) and is again evident in the twelfth century (for example, the *Natural Questions* by Adelard of Bath), it received a unique format in the course of the thirteenth and subsequently became the virtual embodiment of scholastic science during the fourteenth, fifteenth, and sixteenth centuries. While some of the questions varied, many were in vogue for centuries. A typical set of questions appears in Book IV of the *Questions on the Eight Books of Aristotle's Physics* by Albert of Saxony (*ca.* 1316–1390), a famous scholastic commentator at the University of Paris. Among the seventeen questions considered, we find:

1. Whether place is a surface.

3. Whether place is immobile.
5. Whether the natural and proper place of earth is in water or inside the concave surface of water.
6. Whether the concave [surface] of the moon is the natural place of fire.
8. Whether the existence of a vacuum is possible.
9. Whether, in its downward motion, a heavy simple body has an internal resistance; and, similarly, [whether], in its upward motion, a light [simple] body [has an internal resistance].
10. Whether a resisting medium is required in every motion of heavy and light bodies.
11. Whether, if a vacuum existed, a heavy body could move in it.
12. Whether something could be moved in a vacuum—if one existed —with a finite velocity or local motion or motion of alteration.
13. Whether condensation and rarefaction are possible.

In their responses to such questions, scholastic commentators adhered to a formal procedure that had evolved from verbal disputations held at the universities. The enunciation of the question was always followed by one or more solutions supporting either the affirmative or negative position. If the affirmative position was initially favored, the reader could confidently assume that the author would ultimately adopt the negative position; or conversely, if the negative side appeared first, it could be assumed that the author would subsequently adopt and defend the affirmative side. These initial opinions, which would subsequently be rejected, were called the "principal arguments" (*rationes principales*). The author might then describe his mode of procedure and perhaps further clarify and qualify the question, or define and explain particular terms in it. He was now ready to present his own opinions, usually by way of one or more elaborated conclusions or propositions. Occasionally, in order to anticipate objections, he would raise doubts about his own conclusions and subsequently resolve them. At the very end, he would respond to each of the "principal arguments," thus terminating the question by formal rejection of opposing viewpoints. Although it is true that the large majority of these *Questiones* were unimaginative and repetitive, it is more noteworthy that a number of eminent schoolmen (for example, William of Ockham, John Buridan, Nicole Oresme, Albert of Saxony, and others) found this austere literary format a convenient vehicle for interpreting Aristotle and initiating some of the major departures from his physics and cosmology.

As central as the works of Aristotle were to medieval education and intellectual life, his scientific and philosophic works were viewed by theologians with suspicion and hostility through much of the thirteenth century. Indeed, in the latter part of that century, a theological reaction to Aristotle and his zealous followers produced consequences that profoundly affected the course of medieval philosophy and, though more difficult to detect, appears to have influenced the character and substance of scientific discussion. Fear of Aristotle's influence stemmed from his books on natural philosophy, which contained judgments and opinions that were subversive of Christian faith and dogma. The most offensive of his definitive conclusions were as follows. (1) The world was eternal, which effectively denied God's creative act. (2) An accident or property could not exist apart from a material substance, a view that clashed with the doctrine of the Eucharist. According to this doctrine, after the whole substance of the bread and wine has been changed into the whole substance of the body and blood of Christ, the visible accidents of the bread and wine continue to exist though not inhering in any substance. (3) The processes of nature were regular and unalterable, which eliminated miracles. (4) And, finally, the soul did not survive the body, which denied the fundamental Christian belief in the immortality of the soul. Moreover, by denying Plato's theory of ideas and creation in time, Aristotle's philosophy effectively denied the Augustinian doctrine of exemplarism in which, through all eternity, God is said to have known all species of things that He would eventually create.

Trouble was not long in coming. In 1210, soon after Aristotle's works in natural philosophy had become available in Latin, the provincial synod of Sens decreed, under penalty of excommunication, that the books of Aristotle on natural philosophy and all commentaries thereon were not to be read at Paris in public or secret. Confined to the locale of Paris, this ban was repeated in 1215 specifically for the University of Paris. On April 13, 1231, the same ban was modified and given papal sanction by Pope Gregory IX, who, in a famous bull, *Parens scientiarum*, (often called, for other reasons, the Magna Carta of the University of Paris), ordered the offensive Aristotelian treatises purged of error, for which purpose he appointed a three man commission on April 23. For reasons that are as yet unknown, the Pope's committee failed to submit a report and the command to expurgate the books of Aristotle was never executed, although, curiously, in 1245, Pope Innocent IV extended the ban to the University of Toulouse, from whence had emanated, some years earlier (1229), an invitation to masters and students to come to Toulouse where the books of Aristotle, forbidden at Paris, were openly studied and heard. On admittedly scanty evidence, it seems that the ban at Paris was in effect for approximately forty years, since it appears that only Aristotle's ethical and logical works were publicly taught there (the physical and philosophical works were undoubtedly read privately) until 1255, at the latest. In that year a list of texts in use for lecture courses at the University of Paris included all of Aristotle's available works. Thus the onerous, but impractical restrictions placed upon Parisian scholars were at an end and they could now enjoy the same privileges as Oxford scholars, who had never been denied the right to study and comment upon *all* the works of Aristotle during the long years of prohibition at Paris.

With the way now open for legitimate public study, students and teachers in the arts debated and discussed Aristotelian natural philosophy and metaphysics, and applied its modes of philosophical analysis to resolve problems in all areas of human thought. As a guide to the genuine thought of Aristotle, many followed the Aristotelian commentaries of Averroes, whom they called "the Commentator," as a mark of deep respect, just as they called Aristotle "the Philosopher." And just as Averroes before them had kept philosophy and theology distinct, so also did many of his Christian followers. Rather than seek to reconcile Aristotle's philosophy with Christian faith, they sought to show instead that according to Aristotle and philosophy the eternity of the world, unalterable regularity of natural events, and other doctrines at variance with standard thirteenth century theological interpretations, were either demonstrable by natural reason or, at the very least, could not be proven false. When pressed, some of these masters would readily have agreed that where Aristotle's opinions conflicted with faith, the latter must be followed. For example, Siger of Brabant (*d. ca.* 1285), a Parisian master of arts, insisted that it was his intent to explain Aristotle's views independently of revealed truth. Certain philosophical opinions that were in opposition to faith could not, he argued, be disproved by natural reason. Nevertheless, he adhered to the truths of the faith. Another contemporary, Boethius of Dacia (or Sweden), ignored revelation as lying beyond reason; he devoted himself solely to philosophy, by means of which he subjected all arguments to human reason. Boethius insisted that the philosopher must use only natural reason in his inquiries about the physical world and its principles. On this approach, it was right to argue that one could not demonstrate the temporal beginning of the world or the resurrection of the dead, for these are strictly matters of faith. Although Siger and Boethius did not explicitly endorse a doctrine of "double-truth," wherein a proposition in philosophy might be true in the natural domain and its contradictory independently true in the realm of faith, they represented a trend of thought which left the theologians disturbed and uneasy. The theologians, who had themselves employed Aristotelian philosophical language, concepts, and arguments to elucidate and explain theological doctrines, took little comfort from repeated disclaimers by philosophers that despite philosophic demonstrations contrary or inimical to the tenets of faith, the latter took precedence over all else. Thomas Aquinas voiced displeasure at the philosophers who declared that in presenting certain philosophical ideas contrary to faith, they were merely expounding the views of Aristotle to which they themselves did not subscribe.

The tension that developed at the University of Paris between the masters of arts, who were teachers of philosophy but untrained in theology, and the theologians put them on a collision course. They were divided by a fundamental issue. If the principles of natural philosophy are necessarily true, they must conflict directly with revealed religious truth. But if they are only probable, natural philosophy, that is, the science of nature in the broadest sense, is not demonstrative and cannot possibly arrive at certain truth.

In 1267, Bonaventure denounced the masters of arts who believed in the eternity of the world, a single intellect for all men, and the impossibility of attaining immortality. A few years later, in 1270, the bishop of Paris intervened and condemned thirteen propositions derived directly from the teachings of Aristotle or through the interpretations of Averroes. It was now punishable by excommunication to accept as true the eternity of the world, a single intellect for all men, the necessary control of terrestrial events by celestial bodies, that God does not know beings other than Himself, and so on. In 1272, the masters of arts at the University of Paris were required by oath to avoid consideration of theological questions, but if for any reason, they found it unavoidable, they were sworn further to resolve all such questions in favor of the faith. The intensity of the controversy was underscored by Giles of Rome's *Errors of the Philosophers* written sometime between 1270 and 1274. In this treatise, Giles compiled a list of errors drawn from the works of the non-Christian philosophers, Aristotle, Averroes, Avicenna, Algazali, Alkindi, and Moses Maimonides.

These lists of errors were but a prelude to the crashing climax that came in 1277, when Pope John XXI, concerned over the divisive intellectual unrest, instructed the bishop of Paris, Étienne Tempier, to investigate the controversies besetting the University of Paris. Investigate he did. Within three weeks, and on the advice of the theologians, he issued a blanket condemnation of 219 propositions drawn from many sources, including the works of Thomas Aquinas, who had died three years before. Though Tempier acted without the knowledge of the Pope, the latter acquiesced in the action which set excommunication as the penalty for all who held even one of the damned errors. In a scathing preamble to the list of condemned propositions, the bishop of Paris denounced the faculty members at the University of Paris who treated certain errors contrary to faith as if they were merely doubtful rather than abominable falsehoods. Equally heinous, in the bishop's view, was the apparent tendency of some to follow the doctrine of the double truth. Although blatant proponents of a doctrine of double truth have yet to be identified, the bishop of Paris was led to believe that in the arts faculty there were men who would dare maintain that two contrary truths could be defended if it were only argued that one was true if demonstrated by natural reason from the premises of physics and metaphysics (though it might be false in theology and faith) and the other true by virtue of faith and dogma (though false by natural reason).

The large number of condemned propositions, which included the thirteen denounced in 1270, formed a strange mixture of beliefs and opinions some of which were probably never expressed publicly in writing but may have been uttered orally in public disputes or private conversation. A few of the articles enable us to understand the sense of outrage and hostility which the philosophers must have aroused in the theologians if the following assertions had wide currency:

152. That theological discussions are based on fables.
153. That nothing is known better because of knowing theology.
154. That the only wise men of the world are philosophers.

The doctrine of the double truth finds expression in article 90, which says "That a natural philosopher ought to deny absolutely the newness [that is, the creation] of the world because he depends on natural causes and natural reasons. The faithful, however, can deny the eternity of the world because they depend upon supernatural causes." Among the propositions derived from Averroes we find not only the eternity of the world (article 87), but also the claims "that there was no first man, nor will there be a last; on the contrary there always was and always will be generation of man from man" (article 9) and "That to make an accident exist without a subject is an impossible argument implying a contradiction" (article 140).

A number of propositions were offensive because they were deterministic and placed limits on the power of God to act freely and unpredictably. Among these, the following are noteworthy (as will be seen, articles 34 and 49 were destined to play a significant role in fourteenth century scientific discussions):

21. That nothing happens by chance, but all things occur from necessity and that all future things that will be will be of necessity, and those that will not be it is impossible for them to be. . . .

34. That the first cause [that is, God] could not make several worlds.

35. That without a proper agent, as a father and a man, a man could not be made by God [alone].

48. That God cannot be the cause of a new act [or thing], nor can He produce something new.

49. That God could not move the heavens [that is, the sky and therefore the world] with rectilinear motion; and the reason is that a vacuum would remain.

141. That God cannot make an accident exist without a subject, nor make more [than three] dimensions exist simultaneously.

147. That the absolutely impossible cannot be done by God or another agent—an error if impossible is understood according to nature.

Since automatic excommunication was the penalty for anyone bold enough to defend as true any of the 219 propositions, the impact of the condemnation of 1277 was immediate and lasting. It set in motion a vigorous assault by theologians against the broad claims of philosophers, and even certain theologians, concerning the alleged demonstrability of certain theological doctrines and dogmas. The restrictions placed upon God's absolute power by Aristotelian philosophers and some theologians were gradually removed by showing the inconclusiveness of certain philosophical proofs in the realm of theology. Ultimately, the only basis for accepting theological dogmas and beliefs was faith alone. In the process of defending the absolute power of God to act as He pleased, the theologians not only humbled philosophy and philosophers, but used philosophical arguments to show how futile were attempts to demonstrate what God could or could not do, or to prove His existence or attributes. This was achieved only after a searching critique of knowledge that led eventually to philosophical empiricism and the nominalism of the fourteenth century. Although William of Ockham (*ca.* 1280–*ca.* 1349) did not initiate this process, he brought it to fruition. What ecclesiastical decree and the threat of excommunication had sought to accomplish in 1277, was finally achieved by philosophical argument during the first half of the fourteenth century.

A gifted logician and philosopher, Ockham was above all a profound theologian. In his view, the world was utterly dependent on the unfathomable will of God, who, by His absolute power, could have made things other than they are. From this it followed that all existent things are contingent—that is, they could have been made otherwise or not at all. As a completely free agent, God can do anything that does not involve a contradiction. Whatever He could create through secondary or natural causes, He could also produce and conserve directly without intermediate agents. So great is God's power that He could, if He wished, create an accident without its substance, or a substance without its accidents; or produce matter without form, or form without matter. From these strictly theological considerations, which so thoroughly reflected the theological spirit that gave rise to the condemnation of 1277, Ockham was led to an epistemology that has been characterized as radical empiricism.

The fundamental feature underlying his empiricism was the conviction that all knowledge is gained from experience through "intuitive cognition," an expression which Ockham adopted from Duns Scotus. By this Ockham meant that objects external to the mind, as well as personal mental states, were apprehended directly and immediately. From such direct apprehension, one could know whether or not something existed. No demonstration was required, and none could be produced, to show the existence of anything apprehended in this manner. Indeed, even if an object were lacking or inaccessible, it might yet produce an intuitive cognition, since God Himself might choose to supply the cause of the cognition directly rather than operate in the customary manner through a secondary cause. In either case our experience of that object would be the same. Furthermore, God could also produce in us the belief that a nonexistent object actually existed. Psychological certitude was thus rendered indistinguishable from certitude based on "objective" evidence acquired through the senses. Although Ockham could hardly be classified as a sceptic, his failure to provide appropriate criteria to distinguish between these cases (for example, criteria of a kind that might enable one to differentiate between a dream or hallucination and reality) could easily generate scepticism, and may have done so among some of his followers.

The consequences which Ockham drew from this empiricism were truly "radical." He argued that knowledge of one existent thing does not permit us to infer the existence of any other thing, since necessary connections cannot be assumed between contingent things, which, apart from God, are the only kinds of existents. Therefore, *a fortiori,* he could see no justification for inferences from experience to what transcends experience. Hence it would be futile to attempt to demonstrate God's existence from the order of natural existence as perceived through the senses. Since Thomas Aquinas' well-known proof for the existence of God was of this kind, it was rejected as unwarranted (Ockham offered a very different kind of proof for the existence of God, but denied the validity of demonstrations which sought to prove that of necessity no more than one God could exist).

By denying necessary connections between contingent things, Ockham was led on to a critical examination of the fundamental notion of causal relations. In his *Sentence Commentary,* he held that something could be construed as an immediate cause when the effect it produces occurs in its presence and—all other conditions being equal—fails to occur in its absence. Only by experience, however, not a priori reasoning, could sequences of events meeting the conditions described here be justifiably characterized as causally related, as, for example, when we detect that fire is the cause of combustion in a cloth. A priori reasoning could play no role, as it did in earlier discussions of causality, because Ockham had shown that the existence of one thing could not necessarily imply the existence of another thing. But even experience could not provide certainty in establishing causal relations, since God might have dispensed with the secondary cause and set fire to the cloth directly. Criteria for distinguishing between the direct act of a secondary causal agent and the direct intervention of God as primary agent were not supplied by Ockham. Even under the most ideal conditions for repeated observation of sequences of events, it would be impossible to identify the particular causal agent with certainty. This state of affairs seriously undermined the Aristotelian sense of definitely knowable and necessary cause–effect relationships that had been so widely and, perhaps, uncritically accepted in the thirteenth century.

Not even Ockham's own famed "razor," or principle of economy, could be applied to the radically contingent world that he envisioned. For it was not possible to determine by any cognitive means whether the world in which we live is simple or complex. God could willfully choose to produce events and things in ways that are complicated rather than simple. Indeed, the human mind was not even capable of determining whether the world was a teleological creation. Thus the cutting edge of the razor was reserved solely for philosophical entities and descriptions used to explain the world and its processes. Here the simplest explanation was always to be preferred. With this as his objective, Ockham cut away a host of medieval "forms" previously invoked as explanatory devices or as formal realities in philosophy and theology.

As a thoroughgoing empiricist, Ockham insisted that the basic premises of science be derived from experience. Since such premises could not involve necessary cause–effect relations, they had to be expressed as conditional or hypothetical statements. Existence of the things represented in the conditional propositions could not be inferred with necessity, nor could any connection be claimed between such propositions and the actually existing contingent world. This attitude, in conjunction with the influence of the notion of God's absolute power, and a general nominalist conviction that what is not observable is not real, was reflected sometime later by John Buridan (*ca.* 1300–*ca.* 1358). In discussing Aristotle's rules of motion in his *Physics Commentary,* Buridan remarks that in the formulation of these rules moving forces are assumed constant, although no such forces have been observed in nature. "And from these things it seems to me it must be inferred that these rules are rarely, or never, found to produce their effect. Nonetheless, these rules are conditional and true for if the conditions stated in the rules were observed, everything would occur just as the rules assert. For this reason it ought not to be said that the rules are useless and fictitious because although these conditions are not fulfilled by natural powers, it is nevertheless possible, in an absolute sense, for them to be fulfilled by the divine power."

Ockham's influence on the intellectual currents of the fourteenth century was profound and lasting. It produced a widespread tendency to accept empiricism as the foundation of whatever true knowledge was attainable. Unobservable entities and forms were deemed unreal and lost their status as ontological explanatory devices. Empiricism and rejection of the reality of the unobservable came to be characteristic features of the nominalist mode of thought in science and philosophy.

Among Ockham's followers who were theologians, there was a marked inclination to uphold, and pursue even further, the consequences of his critique of knowledge which had sought to narrow drastically the applicability of philosophical proof to the domain of theology. In this group, which included John of Mirecourt, Pierre de Ceffons, Robert Holcot, Jacques d'Eltville, Pierre d'Ailly, and others, Nicholas of Autrecourt was the most prominent and perhaps the most radical. Nicholas, who has been called the medieval Hume, insisted that certitude of evidence could have no degrees. Arguing from the Ockhamist position that the existence of one thing does not allow us to infer the necessary existence of any other thing and that a cause cannot be logically inferred from its effect, Nicholas denied all the propositions of Aristotelian natural philosophy. In his view the multitude of conclusions in Aristotle's natural philosophy were indemonstrable and therefore inconclusive. He concluded that only probable knowledge was possible and sought to present a more probable alternative to Aristotelian physics that was based upon the very Greek atomism vigorously rejected by Aristotle. Nicholas argued that motion and change could be explained more plausibly by the movement of invisible and indivisible atoms.

If the Ockhamist theologians tended toward radical interpretations, the same may not be said of Ockhamist masters of arts. The latter, who were not formally trained in theology, were generally forbidden to discuss theological questions. Of perhaps greater significance was the fact that their professional careers were devoted to the teaching and study of Aristotelian natural philosophy. In contrast to the theologians, who found Aristotelian philosophy with its "demonstrations" a threat to theology, faith, and the absolute power of God, the masters of arts had a vested interest in preserving a reasonably high degree of respectability and plausibility for Aristotelian explanations in science and philosophy. Hence, although the critical tendencies in fourteenth century thought also produced in them, as we shall see in later chapters, a critical attitude toward Aristotle, they were far from wanting to undermine the foundations of his philosophy and world view. Eminent fourteenth century Parisian masters like John Buridan, Albert of Saxony, and Marsilius of Inghen, accepted the emphasis on empiricism (after all, Aristotle, in opposition to Plato, had stressed the empirical foundations of knowledge, although in a manner different than Ockham) but argued, against theologians such as Nicholas of Autrecourt, that knowledge acquired inductively from observation and experience could provide a certitude that was wholly adequate for the requirements of natural science. Rather than emphasize the uncertainty of our knowledge of causal connection, they stressed the positive aspects of human knowledge. For them incomplete induction could produce a degree of certainty entirely adequate for rendering judgments in natural science. Buridan, for example, argued that if by experience we fail to discover an instance in which fire is not hot, the general claim that every fire is hot is warranted. On the basis of such empirical evidence, he rejected the existence of vacua in nature arguing that everywhere in nature we experience material bodies. Indeed another kind of regular experience only serves to reinforce this conclusion, for, without exception we are aware that in separating two bodies another body always intervenes. Against Autrecourt's rejection of all indemonstrable principles, Buridan argues that many such principles must be accepted without demonstration and on the basis of induction by means of the senses, memory, and experience, as, for example, that fire is hot and that the sun produces heat.

Thus for Buridan, as for many others, general principles arrived at through induction were a proper foundation for the pursuit of science. Where many theologians tended to employ Ockham's radical empiricism to undermine science and reduce its conclusions to probabilities, arts masters seized upon empiricism as an acceptable foundation for the sciences whose fundamental principles, though admittedly indemonstrable, had all the truth they needed.

Despite a difference in attitude toward the fundamental principles and foundations of science, both theologians and arts masters spoke frequently of "saving the appearances" or "phenomena." By this was meant either that different hypotheses or explanations could equally well save a particular physical phenomenon; or that one explanation might seem more plausible than other alternatives. In such cases, physical reality was not claimed for the mechanisms of explanation. Formulated by the ancient Greeks, it was first employed in astronomy where the motions of the planets were represented by means of circles and combinations of circles, or eccentrics and epicycles, as they were called. The Greeks sought to represent the observed planetary motions as accurately as possible and, for the most part, cautiously avoided the claim that eccentrics and epicycles reflected the true motions of the planets. In a passage that was known from 1271 on in a Latin translation, Simplicius, an important Greek commentator of the 6th century A.D., mentioned in his commentary on Aristotle's book *On the Heavens* that certain astronomers adopted different hypotheses about the mechanisms of celestial motion without declaring that such mechanisms actually existed in the heavens. Nominalistic tendencies and influences in the fourteenth century only served to reinforce the doctrine of "saving the phenomena" and probably provided the philosophical justification for its extended application to physics and natural philosophy generally. Eminent fourteenth century scholastics such as John Buridan, Nicole Oresme, Albert of Saxony, Pierre d'Ailly, and many others, found such explanations convenient and useful.

But the uncertainty generated by nominalism did more than merely reinforce and encourage the formulation of probable and plausible explanations to save physical phenomena. It was perhaps instrumental in prompting some scholastics to avoid commitment to the question of the validity of scientific principles and the knowability of causal connections. In their scientific discussions contingent physical phenomena were considered in hypothetical form. By emphasizing logical rigor and making no claim about existential implication, Ockham may have encouraged a tendency at Oxford and Paris to imagine all manner of possibilities—and even seeming absurdities—without regard for physical reality or application. The characteristic sign of this approach was the phrase *secundum imaginationem*—"according to the imagination." Nowhere was this tendency more apparent than in treatises and discussions concerned with the ways in which the intensities of qualities or motions might be conceived to vary. In the hypothetical and imaginary physical problems formulated under these circumstances, unobservables and formal distinctions could be introduced almost at will because the conclusions reached were not claimed to be a reflection of physical reality, or applicable to it.

In our sketch of the major intellectual consequences which flowed from the condemnation of 1277, we have yet to ask how, if at all, the condemnation actually influenced the course of medieval science. By undermining confidence in the deterministic claims of Aristotelian natural philosophy, did it free medieval science from bondage to Aristotle's cosmological and philosophical prejudices and modes of argument? Were articles 34 and 49, which compelled all to concede that God could move the universe in a straight line even if a vacuum were left and to allow that He could create as many worlds as He pleased, sufficiently subversive of Aristotelian cosmology and physics to stimulate scientific imagination in fruitful but non-Aristotelian directions? Did they serve to give birth to modern science, as Pierre Duhem, the great pioneer investigator into the history of medieval science, claimed? If they did, it would be ironic that an infringement of freedom of expression and inquiry should have brought forth modern science.

Were this interpretation substantiated, it would inevitably suggest that the scientific revolution, usually associated in its beginnings with the great name of Galileo, was but the continuation of anti-Aristotelian scientific currents generated in the fourteenth century. Galileo's repudiation of Aristotelian physics could then be viewed as the culmination of intellectual forces steadily and inexorably at work since the 1270s.

Or does the truth lie elsewhere? Was the condemnation of 1277 productive of little substantive change in the fabric of Aristotelian science, as Alexander Koyré, perhaps the most eminent historian of the scientific revolution, would have it? Were the condemned cosmological articles, made so much of by Duhem, merely of nuisance value? Did they serve only to hamstring and bedevil the pursuit of Aristotelian physics and cosmology by extracting forced pronouncements that God, by His absolute power, could have done or made things differently? Were the alternatives cited by the arts masters viewed as genuine options to Aristotelian conclusions or only distasteful possibilities enunciated under duress? Was the emphasis on God's absolute power to produce any physical act short of a logical contradiction detrimental to the intensive pursuit of Aristotelian science, a science whose parts were too integrated to really permit its adaptation to the theological requirements of the condemnation? Indeed, if the condemnation was efficacious in generating a radical reaction to Aristotelian science, as Duhem believed, why was medieval Aristotelian science not altered more drastically, or better yet overthrown, in the fourteenth and fifteenth centuries? Why was its total repudiation delayed until the late sixteenth and early seventeenth centuries?

No wholly satisfactory resolution of these significant questions has yet appeared. A thorough study of the impact of the condemnation on scientific development and achievement would greatly aid in formulating appropriate responses. That the condemnation had some effect on scientific discussion cannot be denied. For many years after its supposed nullification in 1325, one or more of the condemned articles were frequently cited by such eminent scientific authors as Bradwardine, Buridan, Oresme, and others. In the chapters that follow a few of the condemned articles will be cited in appropriate contexts. Only at the conclusion of this volume will a brief attempt be made to suggest the overall impact made on the substance of medieval science by the condemnation of 1277. In the absence of detailed studies, this attempt will of necessity be highly conjectural.

CHAPTER IV

The Physics of Motion

THE INTELLECTUAL and historical background relevant to the introduction of Aristotelian science and philosophy and the subsequent
impact of that formidable body of knowledge on medieval thought have
been summarized. In this and the following chapter, we shall attempt
to describe and summarize certain aspects of medieval physics and
cosmology as these developed within the broad framework of Aristotelian science.

Until their resolution in the seventeenth century, some of the most
fundamental problems in the history of physics were concerned with,
or derived from, attempts to explain, for example, why a stone that
was thrown into the air first moved upward at a decelerating rate and
then, after appearing to halt momentarily, descended to the earth with
an accelerated motion. It was probably Aristotle who initially conceived
of the problem in these terms and called the upward motion of the
stone *violent* or *unnatural motion,* and the downward leg *natural
motion.* Both types were subsumed under the more general category
of local motion, or change of place or position, which was one of four
general kinds of change which Aristotle distinguished in the terrestrial
region of the universe extending from the earth to the sphere of the
moon. The other three types which formed a part of the study of
terrestrial physics were change of substance (as when a log is converted
to ashes upon burning), change of quality (as when something changes
color), and increase or decrease of a thing's quantity. These three
changes did not occur in the region extending from the lunar sphere
outward to the sphere of the fixed stars, which represented the extreme
limits of the finite spherical universe. Here ordinary sublunar matter
composed of four elements (earth, water, air, and fire) did not exist.

A fifth and divine element, aether, was assumed to fill this vast supra-lunar space and to form the substance of all celestial bodies. Its divinity was evidenced by its immunity to all change except movement. Only natural motion of a special kind occurred in the celestial region, for the planets and stars were carried round eternally on physical spheres turn-ing in uniform circular paths. Thus Aristotle sharply dichotomized the universe into a terrestrial, or sublunar region, and a supralunar, or celestial, region. The types of matter and their behavior differed radi-cally in each. Of the four types of change in Aristotelian physics, the problems associated with local motion were central to the history of physics.

A general solution to these problems in the sixteenth and seventeenth centuries produced a new physics and a scientific revolution which destroyed and replaced the long dominant Aristotelian physics and cosmology. But long before the emergence of the new physics, there was, in the Middle Ages, considerable dissatisfaction with, and criticism of, Aristotle's explanations. Whether medieval criticism is viewed as part of a continuing tradition of anti-Aristotelianism extending from around 1300 to the seventeenth century, or whether it is viewed as radically distinct from the major anti-Aristotelian assaults beginning with Galileo, medieval physics is worthy of study for its own sake and constitutes a significant chapter in the history of science. To render it intelligible, we first provide a brief account of the major conceptual features of Aristotle's physics as it centered on the problem of local motion. Although Aristotle frequently considered the subject of local motion, nowhere in his extant works is there a systematic and compre-hensive treatment of it. The summary which follows is based upon discussions scattered through a number of his works, especially the *Physics* and *On the Heavens*.

Aristotle's division of terrestrial motion into natural and violent prob-ably originated in gross observation. When falling from heights, some bodies, like stones, were seen to move in straight lines toward the center of the earth, which was said to coincide with the geometric center of a spherical universe. Other bodies, such as fire or smoke, always seemed to rise toward the lunar sphere. Since the class of bodies which fell naturally toward the center of the earth were, on the basis of experi-ence, observed to be heavier than those that rose, Aristotle concluded that, when unimpeded, a heavy, or earthy, body moved naturally down-ward in a straight line toward the center of the earth. Thus the center of the earth—or more precisely, the geometric center of the universe—

was the natural place of all heavy bodies. Conversely, light bodies moved naturally upward in a straight line toward the lunar sphere, which was conceived as their natural place. All these natural motions up or down were characterized as accelerated motions.

Aristotle provided important theoretical explanations for these gross observations. From Empedocles and Plato he adopted the view that all things in the sublunar world are constituted from four basic elements, earth, air, water, and fire. Indeed the matter of every terrestrial body was actually considered a compound of varying proportions of all four elements simultaneously. Bodies that fell naturally toward the earth's center did so because their predominant element was heavy; those that rose naturally upward had a predominantly light element. Earth was deemed absolutely heavy because it would fall toward the earth's center whenever it was above the natural place of earth, whether this was in water, air, or the fiery region above air. Fire was conceived as absolutely light, indeed weightless, and, if unimpeded, would always rise from the regions below toward its natural place above air and below the lunar sphere. Water and air were intermediate elements possessing only relative heaviness and lightness. When below its natural place somewhere within the earth, water would naturally rise; but when above its natural place, in air or fire, it would fall. Air, however, would fall when in the natural place of fire, but rise when in earth or water.

Three pairs of opposites played a significant role in Aristotle's interpretation of the structure of the terrestrial, or sublunar, world. These may be schematized as follows:

Concave surface of lunar sphere	Geometric center of universe (or center of the earth)
Up	Down
Absolute lightness (fire)	Absolute heaviness (earth)

These opposites served as virtual boundary conditions for Aristotle's scattered account of the motion of bodies. The left-hand column tells us that an absolutely light body (fire) would naturally rise rectilinearly upward toward the lunar sphere while the right-hand column informs us that an absolutely heavy body would fall naturally downward in a straight line toward the earth's center. Although Aristotle knew that earth was denser than air and water, he would have denied that this in any way explained the fall of a stone through air or water. A stone falls only because it is absolutely heavy. Fire does not rise to its natural place near the surface of the lunar sphere because it is less dense than

earth, water, and air, but rather because it is absolutely light. Indeed, fire does not even possess weight in its own natural place so that if the air below were removed it would not fall or move downward. Looking back over the history of physics, it is no exaggeration to argue that Aristotle's introduction of absolute heaviness and lightness, with the latter equivalent to weightlessness, was a major obstacle to the advance of physics, though Aristotle thought it a significant improvement over Plato and the atomists who had attributed weight to all things and for whom weight was a relative concept.

Although it was "natural" for heavy bodies, say stones, to fall to their natural place on earth when displaced, Aristotle was moved to offer a causal explanation for this phenomenon. He assumed as a fundamental principle that everything capable of motion, whether animate or inanimate, is moved by something else. It followed, therefore, that a mover, or motive power, was theoretically distinguishable from the thing which it moved. Thus in animate objects, say animals, the soul was the mover and the animal's body the thing moved; in celestial, or planetary motion, the mover was a celestial intelligence and the moved thing was the physical orb of the planet. In both cases, mover and thing moved were distinguishable but not physically or spatially separate from one another. In violent and natural motion of inanimate objects, however, mover and moved were physically distinct. The initial mover in violent motion was often easy to identify since it was necessary that it be in direct physical contact with the thing it moved. Thus a man is the mover, or motive power, when he pulls or pushes a heavy load or throws a stone. While admitting that it was hardly obvious, Aristotle identified the primary cause of unobstructed natural motion as the particular agent (called the *generans,* or generator, in the Middle Ages) which had originally produced the body actually in motion. For example, fire produces fire (as when a log is set ablaze) and confers on the new fire all the properties that belong to fire, one of which is the spontaneous ability to rise naturally when unimpeded. Similarly, whatever natural agent produces a stone confers upon it all of its essential properties, including the natural tendency to fall to earth when displaced from its natural place.

But, as already mentioned, Aristotle acknowledged that all bodies accelerate as they approach their natural places. How did he explain this? Was acceleration an essential feature of natural motion conferred upon all bodies by their generators? Or were all natural motions uniform, with acceleration an additional factor requiring causal explanation? On this important question Aristotle had little to say and it remained for later commentators to supply a variety of explanations. In fact, despite his acknowledgement that natural motions are accelerated, Aristotle treated them as if they were uniform, or, at most, average speeds. Moreover, although he identified the *generans* as a kind of remote motive cause in natural motion, Aristotle discussed fall (and rise) as if weight (or lightness) was the immediate cause of a body's natural downward (or upward) uniform speed. All other things being equal, he concluded that velocity is directly proportional to the weight of the body in natural motion and inversely proportional to the density of the medium through which it moved; that the time of its motion is directly proportional to the density of the medium and inversely proportional to its weight. For example, the speed of a body could be doubled either by doubling its weight (but holding the medium constant) or halving the density of the medium (and holding the weight of the body constant). Similarly, the time of motion could be doubled either by doubling the density of the medium (but keeping the weight constant) or by halving the weight of the body (and holding the density of the medium constant).

For violent motion, Aristotle formulated a series of specific rules in which he described the consequences that would follow from the application of a motive force to a resisting object. Although the rules are couched in terms of force, resisting body, distance traversed, and time, rather than directly in terms of velocity, the latter permits a more convenient summary to be made. The velocity of a body in violent motion is inversely proportional to its own resistive power, which is left undefined, and directly proportional to the motive power or applied force. In symbols, $V \propto F/R$, where V is velocity, F is motive force, and R is the total resistance offered to the applied force, including, presumably, the external medium in which the motion occurs as well as the resisting object or body.

To double the velocity, V, the resistance R could be halved and F held constant; or, F doubled and R held constant. To halve V, F could be halved and R held constant; or R doubled and F held constant. In the reduction of a velocity by half, Aristotle recognized that, with respect to R, F might be weakened to the point where it could no longer move R. Under these altered circumstances, he insisted that motion would immediately cease and the rules of motion no longer apply. Aristotle's qualification was disregarded and ignored in the fourteenth century when his law of motion, $V \propto F/R$, was widely rejected. His critics argued that in halving any velocity, F could be successively halved, or R successively doubled, until F became equal to or less than R, at which point it was assumed as self-evident that motion would cease or be incapable of commencing. And yet, mathematically, the Aristotelian law would indicate a positive velocity, since F/R would, at the very least, be representable by a fraction, however small. It seemed that Aristotle's law committed its supporters either to the physically absurd position that any force, however small, could move any resistance, however large, or to a mathematical theory which continued to register velocities mathematically after they were physically impossible to produce. To avoid this, a new mathematical relationship based on geometric proportionality and, in effect, involving exponents, was substituted. To halve a velocity with the new law, it was necessary to take the square root of the ratio F/R, that is, $(F/R)^{1/2}$ would generate half the velocity produced by F/R. Therefore if F is initially greater than R, and motion occurred, F could not become equal to or less than R, since the reduction of V by half was no longer achieved by halving F or doubling R. To reduce a velocity by one third, it was necessary to take the cube root of F/R, that is, $(F/R)^{1/3}$, and so on. Double or triple velocities could be produced by squaring or cubing F/R, that is, $(F/R)^2$ or $(F/R)^3$. No longer was this accomplished by doubling or tripling F alone, or halving or taking the third part of R alone.

Identification of the initial mover in a violent motion seemed fairly straightforward, but the source of power which enabled a body to continue its motion after losing apparent contact with its initial mover was far from obvious. After a stone had been launched or thrown, what kept it in motion? Aristotle believed that the external medium—air in the movement of a stone—was the source of continuous movement. He believed that the original mover not only puts the stone in motion but also activates the air simultaneously. Apparently, the first portion or unit of activated air pushes the stone and simultaneously activates the adjacent, or second, unit of air which moves the stone a bit further. The second unit, in turn, simultaneously activates the next, or third, unit of air and so on. As the process continues, the motive power of the successive units of air gradually diminishes until a unit of air is reached which is incapable of activating the very next unit of air. At this point, the stone begins to fall with its natural downward motion. By this mechanism Aristotle employed the medium as both a motive power and a resistance. For not only was he convinced that the motive force had to be in constant physical contact with the body it moved, but also that a resistant medium was essential to brake and slow the movement, for otherwise motion could be neither finite nor successive. Indeed, it would be instantaneous, which was absurd. It was taken as obvious that resistance to motion increased as the density of the medium increased and decreased as the medium was rarefied. Since an indefinite rarefaction of the medium would result in a proportionate and indefinite increase in speed, Aristotle concluded that if a medium vanished entirely, leaving a vacuum, motion would be instantaneous (or beyond any ratio, as he put it). This, and other absurd consequences which he believed would follow from the actual existence of void space, led him to argue vehemently against the existence of vacuum in any form. The world was necessarily a plenum filled everywhere in the sublunar region with bodies compounded of the four elements and in the spaces beyond filled with an unchangeable divine aether.

The division of local motion into natural and violent and the numerous concepts, arguments, and physical assumptions which clustered around these two contrary motions constituted the basic core of Aristotle's sublunar physics. Long before Aristotelian physical science reached the Latin West in the twelfth and thirteenth centuries, Greek and Arabic commentators had produced a body of literature in which local motion was intensively discussed. Occasionally significant criticisms were raised and certain of Aristotle's opinions called into question, as when John Philoponus, a Greek commentator of the sixth century A.D., disputed the function and role which Aristotle had assigned to the external medium. Not only did he deny the necessity for a resistant medium in local motion, but he also rejected the external medium, especially air, as the agent or cause of violent motion and suggested instead an incorporeal, impressed force. Arab commentators, familiar with the works of some of the Greek commentators, frequently elaborated and added to these ideas, some of which reached medieval Europe in Latin translation. It was in this manner that Averroes transmitted a brief anti-Aristotelian critique by Avempace (the Latinized form of Ibn Bājja), a Spanish Arab (died 1138), who may have been influenced by Philoponus.

In his commentary on Aristotle's *Physics,* Averroes reports that Avempace denied Aristotle's claim that the time of fall of a body is directly proportional to the density, and therefore the resistance, of the external medium through which it fell. Aristotle's claim would be true, Avempace argued, only if the time required to move from one point to another was due solely to the resistive capacity of the intervening medium. On this crucial point, Aristotle himself furnished Avempace with a powerful counterargument. Aristotle had observed that planets and stars, like all terrestrial bodies, do not move instantaneously from one point to another. And yet Aristotle had also insisted that heavenly bodies move effortlessly through a material celestial aether which offers no resistance. It was obvious that differing finite planetary speeds could occur without the active resistance of a medium. Avempace concluded that not only was a resistant medium unessential for the occurrence of motion, but that its sole function was to retard it. Ordinary observable motion was what remained of hypothetically unobstructed motion after subtraction of the retardation due to the medium. The vagueness of Avempace's account made actual determination of observable motion impossible. He suggested no means by which motion in a resistanceless medium or vacuum might be measured. Is it measured by the weight of the body, by its dimensions, by indwelling power, or in some other way? Precisely how does the total resistance of the medium retard natural motion and produce a final velocity in the form of an observable motion? How is such resistance to be measured? Not until the sixteenth century, when Giovanni Battista Benedetti and Galileo adopted a similar anti-Aristotelian position, was a genuine effort made to provide an objective measure for the resistance of a medium.

Soon after the works of Averroes became available in Latin translation, Avempace's critique became widely known and influential, giving rise to further elaboration and controversy. One of the first to consider it was St. Thomas Aquinas (1225–1274). Though he did not mention Avempace by name, his succinct argument against Aristotle and Averroes left no doubt of his pro-Avempace viewpoint. As empirical evidence that motion in a resistanceless medium would be finite, Thomas repeats Avempace's illustration of motion through the celestial aether, an illustration that soon became commonplace. But reason also tells us that motion in a vacuum would be finite and successive, since void space, no less than space filled with matter, is an extended, dimensional magnitude. To move from one distinct point to another, a body must traverse the intervening empty or full space necessitating that the parts of space nearer the starting point be gone through before those further removed.

Whether or not Aquinas originated this argument, it was destined to become a standard medieval justification for the possibility of finite motion in a vacuum. Such motions were now conceivable in purely spatial and temporal terms. Could they also be made plausible in terms of the usual dynamic principles associated with the ordinary motions of physical bodies? If a real body were placed in a vacuum, assuming that one existed, would it rise or fall with a natural motion? If it were hurled violently, could it move with a continuous motion? Although Aristotle had repudiated the possibility of motion in a vacuum and could furnish no guidelines to those who raised such questions, the medieval response was nevertheless formulated with Aristotelian physical principles firmly in mind, especially those which assumed that whatever is moved is moved by a distinct and identifiable entity and that every motion involves the action of a force operating against a resistance.

The solution with regard to natural motion followed upon the introduction, sometime in the late thirteenth or early fourteenth century, of a new concept, *internal resistance*. This was made possible by a new interpretation of the Aristotelian notion of a mixed, or compound, body. Aristotle had distinguished pure elemental bodies (earth, air, water, and fire), which were mere hypothetical abstractions not actually observable in nature, from compound or mixed bodies, which were mixtures, in varying proportions, of all four elements and were the bodies actually observed in nature. In all mixed bodies, Aristotle had held that one of the elements would dominate and determine the natural motion of the body, that is, whether it would naturally rise or fall. While this interpretation remained acceptable to many during the Middle Ages, some came to believe not only that a mixed body could be composed of two, three, or four elements, but also that a predominant element did not determine its natural motion. Rather, the total power of the light elements was now held to be arrayed against the total and oppositely directed power of the heavy elements. If lightness predominated, upward motion would result; if heaviness, downward motion would follow. The light and heavy elements in mixed bodies were conceived as if composed of parts or degrees. A summation of the parts would reveal the predominance of heavy or light motive qualities and thus determine the direction of natural motion. The greater the ratio of heavy to light parts, the greater the downward speed; similarly, the upward speed would increase as the ratio of light to heavy elements increased.

From this it was an easy step to the concept of internal resistance.

Since heavy and light elements must, by their very natures, move in contrary directions, and since the practice had developed of assigning degrees to each of the elements in a compound, someone apparently took a further step and conceived heaviness and lightness as oppositely acting forces or qualities within one and the same mixed body. The quality with the greatest total number of degrees was, therefore, designated as motive force and its opposite as resistance. If, now, two mixed bodies were compared such that in one heaviness exceeds lightness by eight to three and in the other it exceeds by eight to five, it was reasonable to assume that in the same external medium, the body with the fewer degrees of lightness would fall with the greater velocity. This could be explained by the fact that the quicker moving body has less lightness, or internal resistance. If both bodies had equal degrees of lightness and their downward speeds differed, the quicker descending body obviously had more degrees of heaviness. Generally, in a falling body, heaviness would be construed as motive force and lightness as resistance; in a rising body, lightness would be motive force and heaviness resistance.

Since all potentially observable real physical bodies in the sublunar region were mixed bodies, internal resistance could be employed in the explanation of all natural terrestrial motion. But it was most useful in justifying motion in a hypothetical void, for now the essential generally accepted preconditions of such motion were available, namely motive force and resistance. Internal resistance functioned to prevent instantaneous speed in the absence of an external resistant medium. Every mixed body could be conceived as moveable in the void since it carried force and resistance within itself. But what about pure elemental bodies? Although, as we have already noted, such bodies did not exist in nature, scholastics were still concerned about these hypothetical entities and continued to discuss their motion. The fall of pure, unmixed elemental bodies such as air, water, and earth could occur only in material media (as an absolutely light element, fire could not possibly fall through any medium). In an extended vacuum, however, it was generally held that they would fall with infinite speed since all resistance, both internal and external, would be lacking. Obviously, pure elemental bodies could not possess internal resistance in the sense described for mixed bodies. Since no ratio of motive force to resistance was possible, infinite speed would result. Despite an occasional attempt to conjecture the means by which an elemental body could be conceived to move with natural motion in a vacuum, such motion was generally considered dynamically unfeasible. Only the kinematic, or spatial and temporal, argument proposed by Aquinas, and described earlier, could make it conceptually acceptable.

Within the context of medieval physics and restricted to mixed bodies, internal resistance seemed the most reasonable way of justifying natural motion in a void. Once this was established, an interesting and significant result was soon derived. Thomas Bradwardine (*d.* 1349), Albert of Saxony, and others, concluded that two homogeneous bodies of different size and, therefore, weight would fall in a void with equal speed. From the standpoint of Aristotelian physics, where speed is proportional to heaviness or absolute weight so that the heavier a body the greater its velocity, this was a startling conclusion. For it was now proposed that homogeneous bodies of unequal weight would fall with equal velocities, a conclusion made possible by the assumption of material homogeneity. Each unit of matter in every homogeneous mixed body is identical. Therefore every unit of matter has the same ratio of heavy to light elements, that is, the same ratio F/R, of motive force to internal resistance. Although one body may contain more homogeneous units of matter than another and therefore be larger and weigh more, two such bodies will, nonetheless, fall with equal speeds, since speed was held to be governed solely by an intensive factor, in this case the ratio of force to resistance per unit of matter, rather than an extensive factor such as total weight, as Aristotle had proposed.

Confronting the same problem more than two centuries later, Galileo employed a similar approach (in his *De motu,* or *On Motion,* written around 1590) in rejecting Aristotle's explanation of natural fall. Instead of a ratio of force to internal resistance per unit of matter, Galileo relied on weight per unit volume, or specific weight. He argued that homogeneous bodies of unequal size, and therefore unequal weight, would fall with equal speeds in plenum and void, though their respective speeds in the latter would be greater than in the former. He was led to this conclusion by seizing upon effective weight, rather than gross weight, as the ultimate determinant of velocity. But, for Galileo, effective weight was equal to the difference in the specific weight of a body and the medium through which it fell. Hence it was actually a difference in specific weights which determined velocities. The velocity of a falling body may be represented as $V \propto$ *specific weight of body minus specific weight of medium*; the velocity of a rising body as $V \propto$ *specific weight of medium minus specific weight of body.* Hence, in a void, where the specific weight of the medium is zero, a body would fall with a velocity directly proportional to its specific weight, or weight per unit volume. Obviously, if the specific weights of two unequal bodies are equal, they will fall with equal speed in the same medium or in the void. In his most famous work, *Discourses on Two New Sciences* (1638), Galileo extended the scope of his law declaring that all bodies of whatever size and material composition would fall with equal speed in a vacuum, a generalization destined to become an integral part of Newtonian physics.

The similarity of approach and the nearly identical conclusions reached by Galileo and his medieval predecessors are striking, but perhaps coincidental. Although Galileo may have become aware of medieval discussions during his student days, no hard evidence exists to substantiate this possibility. Indeed, where medieval Aristotelians explained direction of motion in terms of absolute lightness and heaviness functioning as motive qualities, Galileo relied on the relation between weight of body and medium. No longer was it necessary to distinguish the behavior of simple or pure elemental bodies from that of mixed bodies. The widely accepted medieval view that only certain bodies (mixed) could fall with finite speed in hypothetical void space, and that all others (pure elemental bodies) could not, became utterly meaningless in the physics of Galileo's *De motu*. From his most fundamental concept of specific weight, Galileo treated *all* bodies alike, regardless of composition, and concluded that all bodies could fall or move in void and plenum. With Galileo, homogeneous Archimedean magnitudes replaced the elemental and heavy mixed bodies of the later Middle Ages. The inappropriateness of this dichotomy for Galileo's analysis of motion and his rejection of absolute heaviness and lightness rendered the concept of internal resistance meaningless. Internal resistance, which in the Middle Ages had been invoked to permit an explanation in dynamic terms for finite motion in the void, depended upon the contrary tendencies of elements distinguishable as light and heavy in a mixed body. In downward motion, heavy and light functioned as motive force and internal resistance, respectively; in upward motion their roles were reversed. Galileo, however, required neither internal nor external resistance for the production of finite speed in a void, where the speed of a falling body would be directly proportional to its specific weight. Force and the external resistance of a medium, where the latter was a factor, were measured objectively by specific weight. Although specific weight proved inadequate as a mode of explanation for natural fall, its use in the late sixteenth century by Galileo, and somewhat earlier in the century by Giovanni Battista Benedetti, represented an improvement over the vague and ill-defined notions of force and resistance which prevailed in the Middle Ages.

Galileo's more consistent and simplified justification and explanation of fall in void and plenum must not, however, obscure the historical fact that he inherited the very idea of the intelligibility of finite motion in a void from a tradition that is directly traceable to the Latin Middle Ages, to Avempace, and beyond to Philoponus. Indeed, Galileo indirectly acknowledged as much. It was from this anti-Aristotelian tradition that he derived the idea that a resistant medium is merely a retarding factor in the fall of bodies, whose real natural motions occur only in a vacuum, albeit a hypothetical one.

By contrast with the lengthy and rather numerous discussions about the possibility of natural motion in a vacuum, the possibility of violent motion in a vacuum was barely considered. The problem was formidable. Neither of the two essentials in violent motion, namely motive force and resistance, seemed to be present in a vacuum. In the absence of physical media such as air or water, no external motive force or resistance could be invoked, as in Aristotle's explanation of violent motion. Lightness and heaviness, which functioned as internal force and resistance in mixed bodies undergoing natural motion, were of little use in accounting for violent motion. A mixed body in which heaviness predominated must, by definition, be moved upward or horizontally in violent motion so that its predominant heaviness could not operate as a motive force. Other than denying the possibility of violent motion in a vacuum, the only reasonable response consistent with late medieval physical principles is embodied in a statement by Nicholas Bonetus (died 1343?), who relates that "in a violent motion some nonpermanent and transient form is impressed in the mobile so that motion in a void is possible as long as this form endures; but when it disappears that motion ceases."

Much that was significant and vital to late medieval physics is packed into this brief statement. The impressed force described here as a "nonpermanent and transient form" was known as "impetus" in the Middle Ages and represented a radical departure from, and an important addition to, Aristotelian physics. Already in late antiquity, John Philoponus observed that if air in direct contact with an object could cause and maintain its motion for a time, as Aristotle supposed, then it ought to be possible, for example, to put a stone into motion by merely agitating the air behind it. Since this was obviously contrary to experience, Philoponus rejected air as a motive force and assumed instead that an incorporeal motive force, imparted by an initial mover to a stone or projectile, was the cause that enabled the stone to continue its movement. With the impressed force acting as motive power and the stone, or object, functioning as resistance, the requirements for violent motion were met. Air contributed little or nothing to this process. He concluded that violent motion would occur more readily in a vacuum than in a plenum, since no external resistance could impede the action of the impressed force.

Further elaboration of Philoponus' explanation was made by Islamic authors, to whom the impressed force was known as *mail* (inclination or tendency). One of the major Islamic proponents of the *mail* theory was Avicenna, who conceived it as an instrument of the original motive force capable of continuing its action in the body after the original force was no longer operative. He distinguished three types of *mail*: psychic, natural, and violent. Leaving aside the first, which is not relevant to our discussion, natural and violent mail were intended to furnish causal explanations for the two corresponding types of motion differentiated by Aristotle. According to Avicenna, a body was capable of receiving violent mail in proportion to its weight. This explained, for example, why a small lead ball can be thrown through a greater distance than a piece of light wood or a feather. Ontologically, Avicenna conceived of mail as a permanent quality which would endure in a body indefinitely in the absence of external resistances. From this he concluded that if a body were moved violently in a void, its motion would be of indefinite extent and duration since there would be no reason for it to come to rest, a conclusion Aristotle had also reached (without invoking impressed forces) and for which reason, among others, he rejected the existence of void space. Since experience does not reveal motions of this kind, Avicenna also denied the existence of void space. In the next century, Abū'l Barakāt (*d. ca.* 1164) proposed a different kind of mail, one which was nonpermanent and self-dissipating, the type described by Nicholas Bonetus. Thus even in a void, a body in violent motion would eventually cease movement because of the natural and inevitable exhaustion of its impressed force, a consequence which could not be used as a serious argument against the existence of void. These, and other Islamic arguments about impressed force, would eventually find their counterparts in the Latin West. Whether these ideas were directly transmitted via Latin translations of Arabic works or were independently developed in the Latin West is not yet known.

The theory was already known in the thirteenth century, since a few Latin authors, such as Roger Bacon and Thomas Aquinas, rejected the idea that an incorporeal impressed force could account for the continued violent motion of a body. It was not until the fourteenth century, however, that some form of impressed force theory became popular, especially at Paris. As early as 1323, Franciscus de Marchia proposed one version in which the incorporeal impressed force, or *virtus derelicta* ("the force left behind"), as he called it, was a naturally self-dissipating temporary force that was capable of moving a body contrary to its natural inclination. In this process, air continued to play a subsidiary role. For when the body was set in motion, Franciscus believed that the surrounding air also received an impressed force which enabled it to aid in the motion of the body.

The best formed theory was presented by John Buridan, who is perhaps also responsible for the introduction of the term *impetus* as a technical term for impressed incorporeal force. Buridan conceived of impetus as a motive force transmitted from the initial mover to the body set in motion. The speed and the quantity of matter of a body were taken as measures of the strength of the impetus producing the motion. On the correct assumption that there is more matter in a heavy, dense body than in a lighter, rarer body of the same volume and shape, Buridan explained that if a piece of iron and a piece of wood of identical shape and volume were moved with the same speed, the iron would traverse a greater distance since its greater quantity of matter could receive more impetus and retain it longer against external resistances. Thus it is that Buridan seized upon *quantity of matter* and *speed* as means of determining the measure of impetus, the same quantities which served to define momentum in Newtonian physics, although in the latter momentum is usually conceived as a quantity of motion or a measure of the effect of a body's motion, whereas impetus is a cause of motion. Indeed impetus was envisioned as an internalization of the motive force which Aristotle had made external. It seemed a better means of adhering to Aristotle's own dictum that everything that is moved is moved by another.

Like Avicenna, Buridan ascribed to impetus a quality of permanence and assumed that it would last infinitely unless diminished or corrupted by external resistance. Apparently he reasoned that once a mover imparts impetus to a body and the latter moves off and loses contact with the original motive force, no additional impetus could be produced in the absence of any identifiable cause. Since the initial quantity of impetus would not diminish unless corrupted by resistances acting on the body (this would include not only external resistance, but also the natural tendency of the body to move to its natural place), the impetus would remain constant under ideal, though actually unrealizable conditions. Buridan, therefore, clearly implies that if all resistances to motion could somehow be removed, a body set in motion would move on indefinitely and, presumably, in a straight line at uniform speed. There would be no reason for it to change direction or alter its initial speed since not even its inclination to fall to its natural place would be operative as long as we have assumed that all impediments to its forced motion have been removed. Indeed, while the impetus was actively producing a violent motion away from its natural place, it is doubtful whether its inclination to fall to its natural place would be operative or in any way interfere with the violent motion. Unfortunately, Buridan fails to elaborate this significant potential inertial consequence of impetus theory, probably because the very idea of an indefinite rectilinear motion under the ideal conditions just described would have appeared absurd in a finite Aristotelian world. Had Buridan allowed, or even conceived of, such an indefinite rectilinear motion, he would probably have thought it necessary to devise a mechanism for bringing it to a halt. This potential dilemma was avoided when Buridan denied the possibility of finite, successive motion in a void, although, like others who wrote after the impact of the condemnation of 1277, he conceded that God could produce such motions supernaturally. Buridan might have accepted motion in a hypothetical void, if he had adopted a nonpermanent, or self-dissipating, variety of impetus, as described in the earlier statement from Nicholas Bonetus and accepted for a time by Galileo. With a nonpermanent impetus, motion in a void could only have been finite in duration.

Although the concept of an indefinite, uniform, rectilinear motion, an essential ingredient in the principle of inertia, was incompatible with medieval physics, Buridan's permanent impetus incorporated characteristics and properties from which such a motion could be inferred. Before Newton conceived of inertia as an internal force enabling bodies to resist changes in their states of rest or uniform rectilinear motion (the idea that rest and uniform rectilinear motion are identical states of a body was never suggested in the Middle Ages, where rest and motion were considered as contrary conditions, or states), he had thought of inertia much as Buridan's impetus—namely, as an internal force which, in the absence of external forces or resistances, would cause indefinite rectilinear motion.

If indefinite uniform rectilinear motion caused by an impressed force was unacceptable in medieval physics, indefinite uniform circular motion posed no problem. The continued revolution of a mill wheel after it is no longer actively turned, prompted Buridan to conjecture that, in the absence of corrupting resistances, the wheel could be revolved perpetually by the impetus it received when initially put into motion. As an actual possible instance of indefinite circular motion produced by the action of a constant quantity of impetus, Buridan cited celestial motions. Rather than assume the existence of intelligences as celestial movers, as was common, he suggested that at the world's creation God impressed a fixed amount of impetus into each celestial orb. Since the celestial region was thought devoid of all resistance to motion, the original impetus impressed in each planetary orb should remain constant and produce indefinite circular motion, an opinion Buridan proposed tentatively to avoid possible conflict with the teachings of the theological masters.

We have not yet exhausted Buridan's use of impetus. In a manner strikingly similar to earlier Islamic *mail* theorists, he also applied impetus to explain the acceleration of falling bodies. Throughout the history of physics, up to and including Galileo, the problem of fall was treated in a twofold manner. One aspect of the problem was to explain the cause of fall in general without regard to its admitted acceleration; the second aspect was concerned with its acceleration. We saw earlier that Aristotle suggested the generator of a thing as the cause of its natural fall but, in his actual discussions, placed emphasis on weight as the determinant of a heavy body's uniform downward speed. Its acceleration was virtually ignored. In the medieval Latin West, some authors identified a body's substantial form as the cause of its fall, while others, especially in the fourteenth century, considered the heaviness or weight of a body as the primary cause. To account for acceleration, a second, quite separate, cause was sometimes added. Buridan approached the problem in this way. Since the weight of a body remained constant as it fell, he identified its heaviness, or gravity (*gravitas*), as the cause of its natural uniform fall. After eliminating a few commonly discussed possible causes of acceleration (such as proximity to natural place, rarefaction of the air from the heat produced by the falling body, and the lessening of air resistance as the body descended), Buridan explains acceleration by accumulated increments of impetus. This occurs because the heaviness of a body not only initiates its downward fall, but also produces successive and cumulative increments of impetus, or "accidental heaviness," as it was sometimes called. The successive increments of impetus then generate successive and cumulative increments of velocity, thus producing a continuously accelerated motion. Three elements are distinguishable in the process of fall: (1) the heaviness of the body, W; (2) impetus, I; and (3) velocity, V. Initially, at the end of the first temporal instant, Δt, the heaviness or weight, W, produces an original velocity V. Simultaneously, during the same time interval, the heaviness of the body, which remains constant, produces a quantity of impetus, I, which will be operative during the second instant of time and produce an increment of velocity, ΔV. Thus at the end of the second interval of time, $2\Delta t$, the heaviness and impetus, $W + I$, increase the body's speed to $V + \Delta V$. During the second time interval, $2\Delta t$, a second increment of impetus was generated and added to the first. Therefore, in time interval $3\Delta t$, $W + 2I$ will produce a speed of $V + 2\Delta V$. In the fourth interval, $W + 3I$ will increase the speed to $V + 3\Delta V$, and so on. Buridan's explanation is squarely in the Aristotelian tradition since force is always proportional to velocity, and not to acceleration as would be the case in Newtonian physics. This is obvious because every increment of velocity is preceded by a proportional increment of impetus. Thus if after force $W + 3I$ produced $V + 3\Delta V$, no additional increments of impetus were added, the speed would remain constant at $V + 3\Delta V$ and remain proportional to a now constant force of $W + 3I$. Only if weight or heaviness were taken as a constant motive force directly producing increments of velocity, rather than increments of impetus, could it be argued that Buridan had arrived at something similar to the concept that force is proportional to acceleration. But there is little warrant for this interpretation, since the body's weight must first produce an increment of impetus before a proportional increment of velocity can be generated. The connection between weight as a constant motive force and increase of velocity is at best indirect.

Despite some opposition, impressed force theories exercised a continuing influence into the sixteenth century when Galileo himself became an enthusiastic proponent during his early career at the University of Pisa. In his unpublished treatise *De Motu*, he sought to explain the forced upward motion and subsequent downward acceleration of heavy bodies. As the basis of his explanation, he adopted the idea of a residual force, which he derived from Hipparchus, whose views were described in Simplicius' *Commentary on Aristotle's On the Heavens,* a treatise widely known in the Middle Ages. To this Galileo added the mechanism of a self-dissipating incorporeal impressed force, or impetus, which he probably derived from medieval sources. Initially, the mover imparts an impressed force to a stone that is hurled aloft. As the force diminishes, the body gradually decreases its upward speed until the impressed force is counterbalanced by the weight of the stone, at which moment the stone begins to fall, slowly at first, and then more quickly as the impressed force diminishes and gradually dissipates itself. Acceleration results as the difference between the weight of the stone and the diminishing impressed force continually increases. Thus, on the downward leg of the motion, the impressed force actually functions as a resistance. Theoretically, if the body fell through a sufficiently long distance, all of the impressed force would vanish, at which point the body would fall with uniform speed. Eventually, Galileo abandoned the concept of a self-dissipating impressed force and explained accelerated fall by an impetus that is conserved and cumulative, an explanation that differed little from Buridan's.

The problem of accelerated fall gave rise to a historically significant confusion and ambiguity. A body in accelerated fall appeared to increase its speed the longer it fell and the greater the distance it traversed. So intimately do time and distance seem to be associated with fall that increase in velocity seems plausibly relatable to either one or the other, or both together. Buridan, Albert of Saxony, and Leonardo da Vinci adopted both relationships (Leonardo pronounced them in the very same sentence) without realizing the contradictory nature of their commitments. Even Galileo, as late as 1604, mistakenly assumed that velocity is directly proportional to distance ($V \propto s$) rather than time ($V \propto t$), as he later came to realize. The confusion may derive, in part, from the fact that not until the seventeenth century, did it come to be generally accepted that the motion of a freely falling body was uniformly accelerated. Had this general acceptance occurred in the fourteenth century, it probably would have produced the realization that the velocity of a freely falling body is directly proportional to time. Already in the fourteenth century, certain fundamental definitions and proofs were formulated which, when eventually applied to falling bodies by Galileo, produced the conclusion that velocity is proportional to time and the further consequence that the distance traversed by a freely falling body is directly proportional to the square of its time of fall (that is, $s \propto t^2$).

These contributions grew out of a peculiarly medieval concern with the manner in which qualities varied in intensity. How, for example, did colors change their hues and intensities and water become hot and cold? The study and analysis of such problems came to be designated as "the intension and remission of forms and qualities." During the early fourteenth century at Merton College, Oxford University, a group of English scholars with names like Heytesbury, Dumbleton, and Swineshead came to treat variations in velocity, or local motion, in the same manner as variations in the intensity of a quality. The intensity of a velocity increased with speed no less than the redness of an apple increased with ripening. For the next 300 years, from the fourteenth through sixteenth centuries, the analogy between variable qualities and velocity was a permanent feature of treatises on intension and remission of forms and qualities. Although the study of qualitative variations persisted for a long time, the developments which emerged and proved significant in the history of physics were all formulated in the fourteenth century at the Universities of Oxford and Paris, but only after the initial theological and metaphysical context, within which these problems were originally discussed, had been abandoned or ignored.

The medieval contribution centers on original and correct definitions of uniform speed and uniformly accelerated motion, definitions employed by Galileo and on which he could not improve. At Merton College, and elsewhere, uniform motion was defined as the traversal of equal distances in *any* (or *all*) equal time intervals. Like Galileo, medieval authors cogently added the word "any" in order to avoid the possibility of equal distances being traversed in equal times by non-uniform velocities. In generalizing that equal distances are traversed in any and all time intervals, however small or large, their definition guaranteed uniformity of motion.

Extending the definition of uniform motion to the simplest kind of variable speed, the Mertonians arrived at a precise definition of uniform acceleration as a motion in which an equal increment of velocity is acquired in each of any whatever equal intervals of time, however large or small. They also sought to define the difficult notion of an instantaneous velocity. Lacking the fundamental concept of a limit of a ratio, which was only developed centuries later in the calculus, they defined it in terms of uniform velocity. It was expressed as the distance that would be traversed by a moving point or body if that point or body were moved uniformly over a period of time with the same speed it possessed at the instant in question. Although the definition is hopelessly circular, since it defines "instantaneous velocity" by a uniform speed equal to the very instantaneous velocity that is to be defined, the Mertonians merit praise for recognizing the need for such a concept. Galileo employed it in the same form. Not only did they cope with instantaneous velocity directly, if inadequately, by definition, but they also approached it indirectly through their definitions of uniform and uniformly accelerated motion where velocities in infinitesimally small time intervals are clearly implied.

By an admirable and ingenious use of these definitions, the Merton scholars derived what is known as the mean speed theorem, probably the most outstanding single medieval contribution to the history of physics. In symbols, it may be expressed as $S = \frac{1}{2} V_f t$, where S is the distance traversed, V_f is the final velocity, and t is the time of acceleration. Since the velocity is assumed to be uniformly accelerated, $V_f = at$, where a is uniform acceleration. By substitution, $S = \frac{1}{2} at^2$, the usual formulation for distance traversed by a uniformly accelerated motion. When, instead of starting from rest, the uniform acceleration commences from some particular velocity, V_o, a case frequently discussed, the medieval version is representable as $S = [V_o + (V_f - V_o)/2]t$, or simply $S = V_o t + \frac{1}{2} at^2$, since $V_f - V_o = at$.

The mathematical expressions and symbols used above were nonexistent in the Middle Ages. The concise modern formulations used here were expressed rhetorically in a manner that would seem cumbersome, prolix, and perhaps incomprehensible to present day readers. They explain that a body, or point, commencing uniform acceleration from rest, would traverse a certain distance in a certain time. The claim is then made that if the same body were to move during the same interval of time with a uniform velocity equal to the instantaneous speed acquired at the middle instant of its uniform acceleration, it would traverse an equal distance. Thus a uniformly accelerated motion is equated with a uniform motion, making it possible to express the distance traversed by the former in terms of the distance traversed by the latter. Numerous arithmetic and geometric proofs of this vital theorem were proposed during the fourteenth and fifteenth centuries. Of these, the best known is a geometric proof by Nicole Oresme, formulated around 1350 in a work titled *On the Configurations of Qualities,* easily the most original and comprehensive extant treatment of the intension and remission of qualities.

In Fig. 4, let line AB represent time and let the perpendiculars erected on AB represent the velocity of a body, Z, beginning from rest at B and increasing uniformly to a certain maximum velocity at AC. The totality of velocity intensities contained in triangle CBA was conceived as representing the total distance traversed by Z in moving from B to C along line BC in the total time AB. Let line DE represent the instantaneous velocity which Z acquires at the middle instant of the time as measured along AB. If Z were now moved uniformly with whatever velocity it has at DE, the total distance it will traverse in moving from G to F along line GF in time AB is represented by rectangle $AFGB$. If it can be shown that the area of triangle CBA equals the area of rectangle $AFGB$, it will have been demonstrated that a body accelerated uniformly from rest would traverse the same distance as a body moving during the same time interval at a uniform speed equal to that of the middle instant of the uniformly accelerated motion. That is, $S = \frac{1}{2} V_f t$, the distance traversed by Z with uniform motion, equals $S = \frac{1}{2} at^2$, the distance traversed by Z when it is uniformly accelerated. That the two areas are equal is demonstrated as follows: since $\angle BEG = \angle CEF$ (vertical angles are equal), $\angle BGE = \angle CFE$ (both are right angles), and $GE = EF$ (line DE bisects line GF), triangles EFC and EGB are equal (by Euclid's *Elements,* Book I, Proposition 26). When to each of these equal triangles is added area $BEFA$ to form triangle CBA and rectangle $AFGB$, it is immediately obvious that the areas of triangle CBA and rectangle $AFGB$ are equal.

Oresme's geometric proof and numerous arithmetic proofs of the

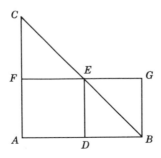

Figure 4

mean speed theorem were widely disseminated in Europe during the fourteenth and fifteenth centuries and were especially popular in Italy. Through printed editions of the late fifteenth and early sixteenth centuries, it is quite likely that Galileo became reasonably familiar with them. He made the mean speed theorem the first proposition of the Third Day of his *Discourses on Two New Sciences* where it served as the foundation of the new science of motion. Not only is Galileo's proof strikingly similar to Oresme's, but the accompanying geometric figure is virtually identical, despite a 90° reorientation, a reorientation that had already been made by some medieval authors.

On some of the most fundamental concepts and theorems pertaining to motion, Galileo had been anticipated by his medieval predecessors. No longer can it be doubted that his contributions in kinematics, once thought to be wholly original, have been grossly exaggerated, largely because the traditional interpretations of Galileo's achievements, formulated between the seventeenth and nineteenth centuries, were developed in almost total ignorance of the achievements of medieval science. Careful and painstaking research in the twentieth century has not only uncovered the hitherto unknown medieval accomplishments, but also taken them into account in reevaluating Galileo's place in the history of science. As a consequence, a new question has been posed. If the major theorems and corollaries formerly attributed to Galileo were already enunciated in the Middle Ages, in what sense, if at all, is it arguable that Galileo founded the modern science of mechanics?

The rectification of the historical record has not diminished Galileo's stature and genius, nor has it deprived him of the right to be honored as the founder of modern mechanics. Although Galileo had been anticipated by some of the medieval contributions described here, and fell heir to others in ways that have yet to be precisely established, his originality and genius derived from an exceptional ability to seize upon and extract what was directly relevant to a mathematical and kinematic description of motion from the diffuse and unfocused medieval doctrine of intension and remission of qualities. The innumerable conclusions and theorems derived in treatises on the intension and remission of qualities and velocities during the fourteenth to sixteenth centuries were little more than intellectual exercises reflecting the subtle imagination and logical acumen of scholastic thinkers. With a minor exception, they were content to treat velocities as variable, intensive qualities divorced from the motion of real bodies. Oresme, for example, characterized geometrical representations of quality variations as fictions of the mind without relevance to nature. Galileo, by contrast, brought together all the significant concepts, definitions, theorems, and corollaries on motion and arranged them into a logical and ordered whole which he applied to the motion of real bodies. Uniform acceleration was no longer a mere definitional concept, but a true description of the way bodies fell in nature. From medieval components he constructed a new science of mechanics which became a vital part of Newtonian science. Such an achievement would alone suffice to rank Galileo with that elite group of scientific geniuses who have, from time to time, profoundly altered the character and direction of science.

CHAPTER V

Earth, Heavens, and Beyond

IN COSMOLOGY, even more than in physics, Aristotle transmitted to the Middle Ages a highly integrated and generally satisfying picture of the structure of the world. Although, from time to time, one aspect or another of it came under severe attack on astronomical, physical, or theological grounds, it remained completely dominant until its overthrow in the sixteenth and seventeenth centuries. It presented a well-ordered, harmonious world that was readily intelligible to most learned men and whose gross features could be vividly and graphically portrayed to all levels of society. Its easy dominance was further facilitated by the fact that during the Middle Ages rival cosmologies from antiquity were either unknown, such as the heliocentric system of Aristarchus of Samos, or known, for the most part, from the hostile works of Aristotle, where they were described in order to be refuted, as with Democritean atomism. Indeed, atomism was doubly damned by its traditional association with atheism, a fact that perhaps explains why the greatest treatise on atomism, Lucretius' *On the Nature of Things,* lay unread through much of the Middle Ages until rediscovered by humanists in the fifteenth century. Moreover, despite a few notable exceptions, Aristotle's cosmology was compatible with Christian scripture and theology.

The most vital concern for Christians, as for Muslims and Jews, was Aristotle's claim, with accompanying "proofs," that the world was eternal, having neither beginning nor end. No act of creation brought the world into being and no act of destruction could end it. A supreme act in the Christian drama was thus denied and, on this point, it was deemed essential to repudiate Aristotle. For a long while, Christians, Muslims, and Jews, sought formally, but without a genuine sense of success, to disprove Aristotle's repugnant claim. Some had naively thought that a proof repudiating the eternity of the world was required to preserve the very foundation of their respective religions. For if it could not be disproved formally doubts might arise as to whether God did, after all, actually create the world. The very basis for belief in God might be seriously undermined and weakened. Moses Maimonides, the twelfth century Jewish philosopher, who lived in the Arab world, saw the danger in all this. If the creation of the world, a fundamental tenet of religion, was doomed to uncertainty unless an impeccable proof demonstrated for all time the impossibility of the world's eternity, the very foundations of religion would totter if none were forthcoming. He argued that neither Aristotle nor anyone else had ever formulated a genuine proof demonstrating the eternity of the universe. Had this been achieved, the teachings of Scripture would have to be abandoned. Readily acknowledging that no successful proof had ever been constructed to demonstrate the divine creation of the world, Maimonides, nevertheless, adopted the latter opinion as indubitable. It had unequivocal scriptural support which he deemed sufficient for its unqualified acceptance. As expressed in his famous treatise, *Guide of the Perplexed,* the approach of Maimonides was known in Latin translation and convinced Thomas Aquinas and others that, since Aristotle's claim for the eternity of the world lacked formal proof, it could be rejected solely on grounds of faith without a formal demonstration of its falsity. And so it was that, although the possible eternity of the world was frequently discussed, its dismissal as contrary to faith, without the necessity of formal demonstration, left medieval Christendom free to accept the rest of Aristotle's cosmology. Curiously, although Plato's cosmology, much of which was available in the Latin translation of Chalcidius, was prima facie more appealing than Aristotle's, for it included the creation of the world by a deity, it was no genuine rival. It lacked Aristotle's elaborate metaphysical foundation as well as the wealth of supporting detail and strong appeals to common sense.

At the geometric center of the vast, though spherical, medieval Aristotelian universe, lay a spherical earth. Contrary to a popular contemporary misconception that prior to the discovery of America by Christopher Columbus the earth was thought to be flat, no flat-earthers of any consequence are known in the Latin West. Aristotle's arguments for a spherical earth were so reasonable and sound that its truth was readily accepted. As observational evidence, he had invoked the curved lines on the moon's surface, inferring rightly that these were cast by the shadow of a spherical earth interposed between sun and moon. He also noted that changes of position on the earth's surface brought different configurations of stars into view, indicating a spherical surface. Even on theoretical grounds, Aristotle was convinced of the earth's sphericity. The fall of heavy bodies in nonparallel lines toward the earth's center indicated the natural formation of a sphere as the innumerable parts of earth accumulated around the center.

Medieval opinions on the comparative size of the earth were drawn largely from Aristotle and Ptolemy, both of whom regarded the earth as but a point in comparison to the vast size of the spherical universe. Repeated by Sacrobosco and accepted by all students who read his famous medieval textbook, *On the Sphere,* this dramatic expression of the comparative vastness of the universe should lay to rest the oft re-peated, though misleading, judgment that the medieval mind took comfort in a small, intimate universe whose coziness was only shat-tered in the seventeenth century with the gradual acceptance of its in-finite extent. If there was a sense of coziness, it derived rather from an assumed intelligibility of the universe than from its size. To the human mind, a universe extending, let us say, for hundreds of millions, or even billions of miles is hardly more imaginable than one that is in-finite, though, of course, it may seem more comprehensible.

Despite the conception of it as a point in relation to the universe, the earth had a measurable size, and at least three estimates were com-monly known. In *On the Heavens,* Aristotle reports that mathematicians estimate the earth's circumference at 400,000 stades, which, despite our ignorance of the exact value of the stade, is a gross overestimation, approximating to twice the actual value of 24,902 English miles. An-other estimate, derived ultimately from Eratosthenes, a Greek of the Third century B.C., was put at 252,000 stades, a good approximation to the true value. It was popularized in the Middle Ages through Sacro-bosco's *On the Sphere* and Pierre d'Ailly's widely known geographical and cosmological treatise, *Image of the World (Ymago Mundi),* com-pleted in 1410 and printed between 1480 and 1485. D'Ailly, who de-rived some of his information from Sacrobosco, describes the alleged method by which the earth's circumference was measured.

Anyone traveling south or north along a meridian until the pole shifted its elevation by one degree would find that he had traversed a distance of 700 stades, which, when multiplied by 360 degrees, yielded a figure of 252,000 stades or 15,750 leagues, where a league equals two miles. According to these figures, the earth's circumference would be

31,500 miles, a considerable improvement over Aristotle but quite wide of the mark. Although Eratosthenes estimated the circumference at 252,000 stades, the method described by d'Ailly is not that used by Eratosthenes, who derived his figure geometrically, but one devised by Arabs in the ninth century by means of which a degree along the meridian was reckoned as 56⅔ miles. From this value of a degree, d'Ailly asserts that Alfraganus, and other Arabs, determined a value of 20,400 miles for the earth's circumference (that is, 360·56⅔), a third estimate considerably less than the true figure.

This last value, conveyed in d'Ailly's *Ymago Mundi* and coupled with a casual remark by Aristotle, played a role in making Columbus' voyage of discovery to America seem plausible. Convinced that he could reach India by sailing westward from Spain, Columbus required evidence to convince Spanish authorities of the feasibility of so bold and costly a venture. He was aware of Aristotle's favorable attitude toward a conjecture that Spain and India were connected by the same ocean. In his personal annotated copy of d'Ailly's *Ymago Mundi,* which is preserved in Seville, Columbus seized every opportunity to emphasize that a degree is only 56⅔ Roman miles and that, consequently, the earth's circumference is 20,400 miles, the smallest of the estimates in popular circulation. Since 56⅔ yielded the smallest value for the earth's circumference and lent credibility to his claim, Columbus expressly declared that it was an exact measurement of a degree, and, in a few instances, drew boxes around this all-important number. A smaller earth made it likely that less ocean intervened between Spain and India. To buttress his case, however, Columbus exaggerated Aristotle's support by attributing to him not only the opinion that a single ocean separated Spain and India, but also the belief that it was navigable in a few days because of its smallness. Far from marking a departure from ancient and medieval opinion, Columbus relied heavily on traditional views to win support for his daring proposal.

Although the earth's central location in traditional cosmology was not seriously challenged until Copernicus proposed his heliocentric system in the sixteenth century, its alleged condition of total rest was carefully reexamined in the fourteenth century. Of the kinds of motion which could be ascribed to the earth, the most significant for the history of science concerned a possible daily axial rotation to explain the risings and settings of all celestial bodies.

The collective authority of Aristotle and Ptolemy guaranteed and almost sanctified the tradition for an immobile earth lying at the center of the universe, with the daily celestial phenomena explained by a daily motion of all celestial spheres. Aristotle explained it by his theory of natural motion and place. The center of the universe was the natural place of a heavy earth which was incapable of rectilinear or circular motion by natural means. Ptolemy, in his *Almagest,* the most thorough and influential astronomical treatise before the appearance of Copernicus' *On the Revolutions of the Heavenly Spheres* in 1543, amassed an impressive array of arguments against axial rotation of the earth relying heavily on human experience and "common sense." Although Ptolemy recognized that an axial rotation might save, or account for, the celestial motions, it could not, in his judgment, explain physical phenomena observable directly above the earth. If the earth actually rotated eastward on its axis, all objects above the earth's surface, including clouds, would appear to be left behind with a westward movement, a phenomenon contrary to experience. Should the air share the earth's rotatory motion, all things in it would, to a terrestrial observer, appear to be left behind, or seem to move westward, as the earth rotated eastward. Even if all objects possessed a rotatory motion in common with earth and air, they would forever remain in the same relative positions. Since we observe directly that objects in the air alter their relative positions, Ptolemy concluded that the earth does not possess a rotatory motion. In the fourteenth century, these arguments were well known and, as we shall see, brilliantly answered.

Arguments in favor of the earth's diurnal axial rotation with the heavens stationary had already been proposed in antiquity, most notably by Heraclides of Pontus (*ca.* 388–310 B.C.) and Aristarchus of Samos (*ca.* 310–*ca.* 230 B.C.). Their ideas were preserved and perpetuated only indirectly by opponents who mentioned them for the sole purpose of refutation, as Ptolemy did. From such hostile sources, the rival theory entered medieval Europe as a discredited scientific opinion. Following the Latin translations of the twelfth and thirteenth centuries, the fact that the possibility of the earth's axial rotation had been seriously proposed, and even defended, in Greek antiquity became well known to all who studied and taught at the universities, or who read cursorily in astronomy and cosmology. Although it was destined to remain an unacceptable theory in the Middle Ages, it received a surprising degree of support from John Buridan and Nicole Oresme, who discussed the problem with considerable ingenuity in their questions and commentaries on Aristotle's *On the Heavens.*

Buridan, who probably wrote first, believed that the daily motion of the stellar sphere and planets could be saved either by the assumption of a stationary heaven and rotating earth, or the reverse. An option in favor of a daily rotation of the earth from west to east, instead of an east to west daily motion of fixed stars and planets, would, however, require that the planetary spheres continue their respective periodic motions from west to east despite the immobility of the outermost sphere. Only in this way could the planets change their positions relative to each other and the fixed stars. If the earth and planets moved west to east in a daily revolution, the earth would complete its rotation in a natural day, the moon in a month, the sun in a year, and so on. In this way, all daily and periodic astronomical phenomena are explicable just as readily as on the alternative hypothesis. In the context of Aristotle's cosmological mechanisms (see below, pp. 71–72), all the celestial spheres formerly involved in the production of the daily motion would now be assumed at rest, while the remainder, whose motions generate the progressive and retrogressive movements along the zodiac, would function as before.

As Buridan recognized, the problem was basically one of relative motion. Although it appears to us that the earth on which we stand is at rest while the sun is carried round us on its sphere, the reverse might be physically true, since the observed celestial phenomena would remain the same. We would be as unaware of such a terrestrial rotatory motion, Buridan insisted, as a person on a moving ship who passes another ship actually at rest. If the observer on the moving ship imagined himself at rest, the ship actually at rest would appear to be in motion. Similarly, if the sun were truly at rest and the earth rotated, we would perceive the opposite. On strictly astronomical grounds, Buridan was plainly convinced that either hypothesis could save the celestial phenomena. Not even astronomers could resolve the issue and determine the physical truth. They were concerned solely with saving, or explaining, the celestial appearances and could employ whichever alternative seemed most convenient.

It seemed that only nonastronomical criteria and arguments could decide between the alternatives. In favor of a daily axial rotation of the earth was the commonly accepted Aristotelian principle that rest is a nobler state than motion. Would it not, therefore, be more appropriate for celestial bodies, especially the outermost and noblest sphere of fixed stars, to be at rest while the earth, the most ignoble body, rotated? Buridan also stressed the desirability of saving the phenomena by the simplest means possible. In this connection, it seemed better to assume that the relatively small earth turns with the swiftest speed while the uppermost and largest spheres remain at rest. To complete a daily rotation, the earth would require a much smaller daily speed than the vastly greater celestial spheres. In this argument, which was essentially reiterated by Oresme, Copernicus, and Galileo, simplicity and credulity were more adequately satisfied by a rotating earth.

Despite these and other arguments favorable to a daily terrestrial motion, Burdian finally opted for the traditional opinion. In his judgment, the earth's rotation failed to explain why an arrow shot upward vertically falls always to the same spot from which it was projected. For if the earth actually rotated from west to east, it ought to rotate about a league to the east while the arrow was in the air. Consequently, the arrow should fall to the ground about a league to the west. Now a supporter of the earth's rotation might counter by arguing that the air moves along with the rotating earth and carries the arrow with it, thus explaining why the arrow falls to the same place from which it was shot. By virtue of the common rotatory motion shared by earth, air, arrow, and observer, the arrow's actual circular motion would go undetected. Because of certain consequences that ought to follow from his impetus theory, Buridan found this explanation unacceptable. When the arrow is projected, a quantity of impetus is impressed in it that should enable the arrow to resist the lateral motion of the air as it accompanies the earth in rotatory motion. As a consequence of its resistance to the motion of the air, the arrow should lag behind the earth and air and drop noticeably to the west of its launching site. Since this is contrary to experience, he concluded that the earth is at rest. In typical nominalist fashion, Buridan derived a crucial physical—not astronomical —consequence of the earth's motion. He was not so rash, however, as to conclude that his experiential appeal constituted a necessary truth. Such a move would have been at odds with the prevailing scholastic attitude of the fourteenth century. Rather, the earth's immobility seemed more probable than the alternative.

In an even more brilliant discussion, Nicole Oresme arrived at the same conclusion. After presenting a series of impressive reasons for supposing that the earth rotates, it comes as a surprise that the cumulative impact of his own good reasons proved insufficient to lure him from the traditional viewpoint. In response to the argument that in ordinary experience we "see" the planets and stars rise and set, from which we infer that the heavens move, Oresme, like Buridan, appeals to the relative motion of ships. Moreover, if a man were carried round by a daily motion of the heavens and could see the earth in some detail, it would appear to him that the earth moved with a daily motion just as it seems to us that the heavens move with such a motion.

To the claim that if the earth turned from east to west a great wind should blow constantly from the east, Oresme counters that the air rotates with the earth. Since we are carried along on the earth, which is accompanied by the air, our experience would be analogous to a person in an enclosed cabin aboard a moving ship. No wind is generated because the passenger, the air, and the cabin which surrounds them are carried along together within the ship.

Another argument, which Oresme ascribes to Ptolemy, is very similar to Buridan's crucial arrow experience, though Buridan is not mentioned. On the assumption of the earth's rotation, Ptolemy concluded that if a passenger in a boat moving swiftly eastward shot an arrow vertically upward, the arrow would fall to the west considerably behind the boat. Similarly, if someone threw a stone straight upward while the earth rotated quickly from west to east, the stone should fall to the west, far behind the place from which it was thrown. Since it falls to the place from which it was thrown and this effect is not observed, Ptolemy concluded that the earth rests. Arguing from his impetus theory, Buridan, as we saw, concurred with Ptolemy. Since the arrow returns to its original place of projection, Buridan inferred that the earth was immobile. Oresme, however, saw nothing incompatible with the return of the arrow and the rotation of the earth. Once again the motion of a ship is employed to illustrate the major points. The movements in a ship sailing eastward occur exactly as when the ship is at rest. Thus, despite the fact that a man's arm would actually undergo two simultaneous rectilinear motions, vertical and horizontal, if he drew his hand vertically downward in line with the ship's mast, his hand would appear to move only with a vertical motion. If we now assume that the earth, the ambient air, and all sublunar matter rotate daily from west to east, the arrow's return to the place from which it was shot can be explained by reference to its two simultaneous, component motions, namely, vertical and horizontally circular (rather than vertical and horizontally rectilinear, as with the arm's motion). Since the arrow shares the earth's circular motion and turns with it at the same speed whether it lies on the earth's surface or is shot vertically into the air, the arrow will rise directly above the place from which it was shot and fall back to it. To the observer, who also shares the earth's circular motion, the arrow will appear to possess a vertical component of motion only.

Oresme concluded that it is impossible to determine by experience that the heavens have a daily motion but not the earth.

The earth's daily rotation was not, however, merely an equally plausible alternative. Positive nonastronomical virtues could be presented in its favor. A terrestrial rotation from west to east would contribute toward a more harmonious universe. All bodies would move in the same direction in periods that increased from the earth outward to the stars. Thus could be avoided the less pleasant alternative in which contrary simultaneous motions—east to west for the daily motion and west to east for the periodic motions—are ascribed to the heavens. Nature's operations should, whenever possible, be explained by the smallest number of simple operations. Since rest is more noble than motion, this arrangement would leave only God in a supreme state of rest, which is fitting and proper. Also, if the motion of celestial bodies diminishes as their distances from the earth increase, then, not only would the earth's motion be quickest, but the sphere of fixed stars, now deprived of a daily motion, would possess the slowest motion of all, making approximately one revolution in 36,000 years to account for the precession of the equinoxes, as proposed by Greek astronomers. Thus excessive speeds for the largest and most remote celestial spheres are avoided.

A daily rotation of the earth would also have obviated the need to create an invisible and starless ninth sphere whose sole function would have been to produce a daily east to west motion of all the planetary spheres and fixed stars. By assigning a daily rotation to the earth, God, who does nothing in vain, would not only have eliminated the necessity of a ninth sphere but also created a less complicated world.

God's alleged inclination to act in the most direct and simple manner is exploited again to explain the miraculous intervention on behalf of the army of Joshua (Joshua 10: 12–14), when God lengthened the day by commanding the sun to stand still over Gibeon. Since the earth is like a mere point in comparison to the heavens, the same momentous effect could have been achieved with a minimum of disruption by a temporary cessation of the earth's rotation. In view of the greater economy of effort, perhaps God actually performed the miracle in this way.

At the termination of such an array of impressive arguments, it comes as a shock to learn not only that Oresme believes the problem is scientifically indeterminate but that he clings to the traditional opinion. The earth rests at the center of the universe and the heavens move around it with a daily motion. The earth's rotation is contrary to natural reason, or ordinary comprehension, as are some articles of the Christian religion. The latter are accepted by faith, but faith, it seems, is an insufficient reason for abandoning natural reason in the determination of a physical or scientific question. On physical, astronomical, and cosmological grounds, the contrary hypotheses are equally tenable. Science and experience cannot decide between them. Oresme's elaborate and brilliant defense of a hypothesis he would ultimately reject was prompted by an ulterior motive. It was meant to protect the Christian faith from demonstrations based on human reason, experience, and science. If these are impotent to demonstrate conclusively a relatively straightforward scientific question, how much more impotent must they be when rashly employed in vain efforts to demonstrate religious dogmas which are believed and accepted on faith alone.

Using reason to confound reason, Oresme reveals that he is heir to the sceptical and probabilistic tradition which emerged from the struggle between philosophy and theology when theologians used philosophy to confound the philosophers. Oresme, the theologian and scientist, shifts the battleground to the realm of science where he confounds the scientists with science and reason. True knowledge could be had by faith alone. On matters concerning the physical world, Oresme, in imitation of Socrates, gleefully confessed that "I know nothing except that I know nothing."

Although Buridan and Oresme concluded that the earth had no rotatory motion, some of their arguments in favor of rotation turn up in Copernicus' defense of the heliocentric system where the earth is assigned both a daily rotation and an annual motion around the sun. Among these we find the relativity of motion, as illustrated by the movement of ships; that it is better for the earth to complete a daily rotation with a very much smaller velocity than would be required by the vast heavens; that the air shares the daily rotation of the earth; that the motion of rising and falling bodies results from a motion compounded of rectilinear and circular elements; and, finally, that since a state of rest is more noble than motion, it is more appropriate for the ignoble earth to rotate than the nobler heavens. Did Copernicus derive some, or all, of these from Buridan and Oresme, whose works were known in eastern Europe, and perhaps even studied at the University of Cracow in the late fifteenth century when Copernicus was a student there? Other than a striking similarity that somehow seems beyond coincidence, there is no evidence that Copernicus knew these treatises or derived his arguments from medieval sources. At the very least, however, Buridan and Oresme are deserving of our praise and admiration for formulating a series of arguments in behalf of the earth's axial rotation that were deemed worthy of inclusion by Copernicus.

Denying a daily rotation of the earth did not mean, however, that no movement at all was to be ascribed to it. Indeed, in the same question where the earth's rotation was considered, and elsewhere in his *Questions on Aristotle's On the Heavens,* Buridan, and others, allowed for incessant, though slight, rectilinear motions of the entire earth arising from continual shifts of the earth's center of gravity, the single point at which the earth's entire weight was said to be concentrated. If the earth were a homogeneous sphere, its center of gravity and center of magnitude would obviously coincide. But the earth is not homogeneous for it consists of parts of varying density and heaviness. Its center of gravity therefore differs from its center of magnitude. The motion of the earth is the inevitable result of a never-ending effort to bring its center of gravity into coincidence with the geometric center of the universe. According to Buridan, whose opinions on this topic were influential, the process may be said to begin with the many small bits of earth that are carried down by streams and rivers from the mountains to the depths of the seas. In this manner, the elevated regions of the earth become lighter and water-covered regions heavier. As the process of weight redistribution continues, the earth's center of gravity shifts. Since it was widely believed that the earth's center of gravity, not its center of magnitude, seeks to coincide with the geometric center of the universe, it followed that the earth would undergo a slight motion until this coincidence was achieved. As this occurs, previously covered regions of the earth would be pushed up and new lands elevated. Parts of the earth presently submerged near the center of gravity are pushed by degrees toward the surface, eventually rising above the waters and extending toward the heights of previously eroded mountains. As the earth's center of gravity shifts continuously to coincide with the geometric center of the world, the process of dissolution is always counterbalanced by the emergence of previously submerged parts.

The mechanisms and processes described here were also useful in explaining the formation of mountains. As the soft stones weather and erode, they are carried down to the sea. The hard stones that remain are repeatedly elevated by each shift of the earth's center of gravity. Over a long period, these aggregations of hard rocks will form high mountains. Much earlier, Avicenna had supplied a better explanation by utilizing the weathering of soft rock as an accidental cause of mountain formation. The hard rocks that remain after deep valleys have been cut emerge as eminences which eventually become mountains. Of interest is the fact that, unlike Buridan, Avicenna also took into account the role of earthquakes in mountain formation—indeed he considered them more fundamental than weathering. Nevertheless, Buridan's geological theories were widely discussed in the fourteenth century and by Jesuits as late as the seventeenth. And, though he rejected them, Leibniz thought the approach ingenious.

Moving upward from the earth into the supralunar region, scholastic commenators adopted a simplified version of Aristotle's cosmology. Aristotle's system, which was based upon earlier works by Eudoxus of Cnidus and Callippus, consisted of 55 concentric celestial spheres which rotated around the earth's axis running through the center of the universe. In the mathematical system of Callippus, on which Aristotle directly founded his cosmology of concentric spheres, the planet Saturn, for example, was assigned a total of four spheres to account for its motion: one for the daily motion, one for the proper motion along the zodiac or ecliptic, and two for its observed retrograde motions along the zodiac. Aristotle gave a physical interpretation to this mathematical scheme. In order to prevent Saturn's proper zodiacal and retrograde motions from being transmitted to Jupiter, the next lowest sphere, he introduced three unrolling spheres whose function it was to counteract the motion of the three spheres which controlled all but the daily motion. Since the daily motion was common to all planets, each was assigned a special sphere for that purpose. In all, seven spheres, instead of four, were now required for Saturn. Similarly, Aristotle thought it necessary to add counteracting spheres for all the planets except the moon, which was located directly above the sublunar region. Thus Callippus' 33 spheres were increased to 55. Rather than cope with these many nesting spheres, some turning in one direction and others in an opposite direction, a basic simplification was introduced in the Middle Ages. Eight (and occasionally nine or ten) physical concentric spheres (plus additions of a nonastronomic kind, as the Empyrean sphere) replaced Aristotle's 55, one assigned to each of the seven planets with an eighth sphere functioning as the carrier of the fixed stars (see Fig. 5). The planets were conceived as fixed, or embedded, in their respective spheres and carried by them with motions that were preserved and continued by their love of God, who moves them indirectly as the object of that love. In their ardent desire to come as close to God as possible, the divine planets move around and around in circular motion. Bold innovators like John Buridan might suggest that impetus drives the

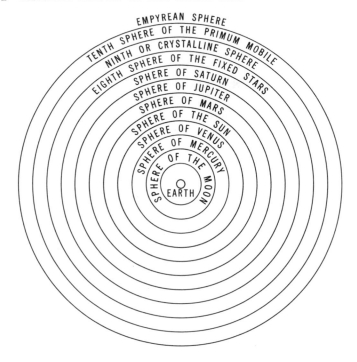

Figure 5

planetary spheres, but most would have agreed with Moses Maimonides that "tis love that makes the world go round."

Astronomically, of course, this system of concentric spheres was even more inadequate than Aristotle's, and it was unable to explain even the obvious variations in planetary distance. To grapple with the technical problems of planetary motion, it was essential to employ the highly developed mathematical astronomy of Ptolemy's *Almagest,* or works derived from it. Among innumerable differences between the Ptolemaic and Aristotelian systems, the Ptolemaic system permitted motions around axes other than those through the center of the universe. This allowed for the development of an astronomy of eccentrics and epicycles (see Fig. 6), which became the very basis of technical astronomy until Kepler's ellipses supplanted them in the seventeenth century. Few in the Middle Ages mastered the intricacies of Ptolemaic astronomy and the scholastic discussions in which epicycles and eccentrics figure were concerned largely with whether or not they had physical existence. Even

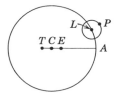

Figure 6   Epicycle on eccentric circle. "*The motion of the center* (L) *of the epicycle that carried the planet* (P) *was regarded as uniform not with respect to the center* (C) *of the deferent [i.e., the circular path traced by* (L)], *or of the earth* (T), *but with respect to another point* (E), *called the 'equant,' i.e.,* ∠ LEA *was considered as increasing uniformly. By properly locating the points* E, C, *and* T, *and determining the ratio of the diameters of the epicycle and its deferent, and choosing proper directions, velocities, and inclinations for the various circles, the apparent irregularities were accounted for.*" (*The figure and accompanying legend are reprinted by permission of the Harvard University Press from Morris R. Cohen and Israel Drabkin,* A Source Book in Greek Science, Harvard University Press, 1958, p. 129. I *have added the bracketed description.*)

so ardent an Aristotelian as Averroes readily conceded that the use of eccentrics and epicycles was appropriate for the computation and prediction of planetary positions, that is, for saving the astronomical phenomena, but vigorously opposed the opinion, held by a few, that epicycles and eccentrics actually had physical existence within the cosmos. Such an admission would have entailed complete abandonment of Aristotelian physics and cosmology by reason of a clear violation of the fundamental tenet that the geometric center of the universe is the only center of celestial motions. For this, and other reasons, Averroes' position was endorsed by almost all scholastics who argued for the probable physical truth of the simplified eight sphere version derived from Aristotle, while granting that Ptolemaic astronomy was essential for saving the astronomical phenomena.

The supralunar region of the medieval cosmos with its succession of tightly nested concentric spheres terminated for some at the eighth sphere of fixed stars and for others at an invisible ninth sphere. Accepting the finitude of the world led inevitably to inquiries about the possibility of existence beyond the outermost celestial sphere. Aristotle had himself raised the question and concluded that all the matter in existence was contained within our finite world. Nothing remained outside from which another world could be formed; nor could any matter come into being there. Since Aristotle's concept of place required that a physical body be capable of occupying it, he inferred that places could not exist outside the world. The nonexistence of material bodies beyond the world also implied the impossibility of a vacuum, which was defined as something in which the presence of a body was potentially possible. Not even time could exist there, since it depended on the motion of bodies. But if neither matter, nor place, nor void, nor time, existed beyond the heavens, then surely nothing whatever lay beyond. It could be described only as utter and absolute privation. Others, including many later Aristotelians, found Aristotle's response narrow and unsatisfactory. They could not refrain from asking what *really* lay beyond the cosmos? Other worlds? Void space? What would be the disposition of a lance, or arm, if it could be pushed through the outermost celestial sphere? Aristotle's response could not repress these ever recurrent questions.

Until 1277, the possibility of other worlds had not been seriously entertained by Christian authors. God's creation was unique and man its focal point. One world was sufficient for the unfolding human drama that moved inexorably toward a great day of judgment. On the issue of one world, Aristotle and his Christian followers were in harmony. But if He wished, could God create other worlds? Prior to 1277, such a hypothetical question would not have been discussed seriously. After 1277, the intellectual context was dramatically altered and the question about other worlds, as well as a host of other unusual and striking questions, was not only raised but its consideration became commonplace. Article 34, which denied that God could make more than one world, was one of the articles condemned in 1277 and its condemnation provoked discussion about other worlds. Thereafter it was necessary to concede that God could have created, or might yet create, worlds other than ours. Seventy-five to one hundred years later authors of such stature as John Buridan, Nicole Oresme, and Albert of Saxony, discussed the possibility of a plurality of worlds with the condemnation of 1277 clearly in mind. Going beyond the mere presentation of new cosmological ideas, they even sought to resolve hypothetical physical problems which posed a serious challenge to Aristotelian physics. Often they extended Aristotelian physical principles to these other hypothetical and possible worlds whose existence Aristotle had categorically denied.

At least three kinds of plurality were distinguished. A succession of single worlds, the version which Empedocles championed and Aristotle had repudiated, did not become the focal point of discussion. It was simply agreed that God could have done this in the past and might yet replace one world by another. A second kind of plurality borders on science fiction and was probably viewed as little more than a mental exercise, hardly to be taken as a serious possibility. Here the plurality is simultaneous but the coexistent worlds are contained one within the other. Oresme insisted that neither experience nor reason could demonstrate the impossibility of such worlds.

The third, and historically significant, version was one that Aristotle had also rejected. It proposed the possibility that separate worlds, lying wholly outside each other, might exist simultaneously in an imaginary space. The possibility of this, to say nothing of plausibility, seemed extremely unlikely in light of the accepted Aristotelian idea that if heavy elements could be placed beyond our world, they would, nonetheless, tend to fall toward the center of the world, since only one true center can exist in the universe. Unimpressed with this explanation, Oresme argued that, with respect to heavy and light objects, "up" and "down" signified only that heavy objects moved naturally to the center of light, or rare, objects. Heavy objects would then be "down" and the surrounding light objects "up." But if a heavy object or piece of earth were beyond our world and separated from it by a vacuum, that heavy object, or piece of earth, would not move toward the center of our world. Without a light body, such as air or fire, to surround the heavy body, up and down are not distinguishable. Hence if void spaces intervened between all coexistent worlds, including our own, the heavy objects associated with each would not tend to move "down" toward the center of our world, since the light objects required to surround the heavy objects, in order to allow a "down" direction to be distinguished, are necessarily lacking in a vacuum. Thus if God created another world like our own, the earth and elements of that world would remain there and behave exactly as their counterparts in our world. To Aristotle's claim that, since all the matter in existence is in our world, no other world could be formed, medieval theologians and philosophers countered that, by virtue of His absolute power, God could create new matter from nothing and create another world.

Despite these, and other, arguments, no actual proponents of a plurality of worlds came forth. It was apparently sufficient to demonstrate that if God did make other worlds—it was, of course, taken for granted that He could if He wished—these would be much like our own, subject to the same physical principles and laws.

Even if no other worlds lay beyond the cosmos, the post-1277 interest in the question of possible extracosmic existence persisted. A positive and significant response was eventually formulated to the hypothetical, but gnawing, question of the disposition of an object pushed through the outermost sphere—a question which had no rationale in Aristotle's austere cosmology. An infinite void was said to lie beyond our material and vacuumless finite world. Because of theological considerations, this proposal was not advanced as a mere possibility but rather as a genuine reality deriving from God's infinite power and omnipresence.

Already in antiquity, certain Stoics, though in agreement with Aristotle that (1) the cosmos itself is a finite sphere without vacua (it was filled with pneuma) and that (2) all existent reality and matter were contained within it, insisted that our finite world was surrounded by an actually existent three-dimensional void capable of receiving matter and serving as its receptacle. The infinite void serves as a receptacle for our finite cosmos. Indeed this seems to be its only function, since interaction between cosmos and infinite void is denied on grounds that the latter has no properties of its own and can in no way affect the material world, which is closed off and sealed from it. The world cannot be dissipated into the void. But granting that the world is surrounded by void, why, it was asked, must it be assumed infinite? In response, it was argued that because no body could exist beyond the physical world, no material substance could limit the void; and since it was absurd to suppose that void could limit void or that void should terminate at one point rather than another—a clear violation of the principle of sufficient reason—the infinity of void space seemed an irresistible conclusion.

Although Stoic cosmology was hardly known in the Middle Ages, one significant argument was made available by Simplicius. In Moerbeke's 1271 Latin translation of Simplicius's commentary on Aristotle's *On the Heavens,* the Stoics are reported to have proved the existence of a vacuum beyond the world. What would happen, they asked, if someone at the outermost extremity of the world extended his arm? Either the arm would reach outside the world from which it could be inferred that a vacuum lay beyond, or the arm would meet an obstacle in the form of matter, in which event the person must then stand at the extremity of the obstacle and again extend his arm. Since the world is assumed finite, this act can be repeated only a finite number of times. Eventually the arm will meet no obstacles at which point a vacuum may be inferred. Although this argument was cited by Aquinas, Buridan, Oresme, and others, its significance was overshadowed by a concept derived from another ancient source.

In an anonymous Latin Hermetic dialogue called *Asclepius,* written in the second or third century A.D., Hermes Trismegistus ("Thrice great Hermes") explains to Asclepius that if void existed beyond the cosmos, which he doubts, it would be void of physical bodies only but never of spiritual substances intelligible to the mind alone. Since *Asclepius* was known in the Middle Ages and Renaissance, it may have transmitted the concept of an extracosmic space filled with spirit but empty of matter, a concept destined to play a vital role in discussions about what might lie beyond the cosmos.

Its influence is detectable on Thomas Bradwardine, who, in his lengthy theological treatise *De causa Dei contra pelagium,* a defense of God against the Pelagians, furnished it with a Christian rationale based on God's infinite power and omnipresence. Assuming that God's perfection would be more complete if He existed in many places simultaneously than in a unique place only, Bradwardine demonstrates that God necessarily exists in every part of the world and also everywhere beyond the real world in an imaginary infinite void. Since God is omnipresent in an infinite void beyond the world, it follows—and here we detect the influence of *Asclepius*—that although void can exist without body, it cannot exist without God's presence.

Theological considerations seem to have impelled Bradwardine to associate these empty places with God. Prior to the creation of the world, a place for it must have existed—indeed an infinity of possible places, since God could have created the world anywhere He pleased. If God had created the place, or possible places, of the world before He created the world itself, the former, not the latter, would have been the first creation and the uniqueness of the creation would be lost. To escape this dilemma, Bradwardine assumed that the infinity of potential places of the world is eternal and uncreated. Co-eternality with God, however, was not to be construed as implying independence from God. Such an interpretation would have signified the existence of two uncreated entities, and greatly diminished God's unique status. In some sense, then, these places and spaces must be associated, and perhaps even identified, with God. How this is to be understood was left unexplained. Are the places in God, or God in the places? Are the places attributes of God? Ignoring these and other questions, Bradwardine sought to avoid more obvious and immediate theological pitfalls while maintaining his conviction that God is omnipresent in an imaginary infinite place void of everything but the deity.

Despite the infinite omnipresence of God, Bradwardine cautions against describing Him as an "infinite magnitude" in the ordinary sense of an extended, dimensional entity. God is infinitely extended only in a metaphysical sense without actual extension and dimension. Incomprehensible as this may seem, it is obvious that, to Aristotle's denial of extracosmic void, Bradwardine formulated a response based on contemporary Christian theology. In so doing, he placed himself in disagreement with such Church luminaries as Augustine and Aquinas, who had supported Aristotle's position. His conclusions were also at variance with the opinions of Duns Scotus (*ca.* 1265–1308) and his followers, who argued that God's will, not his omnipresence, was the basis of divine action. God could act on, and in, a place remote from his actual presence. It was, therefore, unnecessary to assume God's prior presence in the empty place in which He created the world, for which reason Scotus denied the necessity of God's omnipresence in an infinite void space. God's presence in an infinite void beyond the world, as Bradwardine described it, was a new and significant element in medieval cosmology, one that stimulated further discussion in the late Middle Ages and Renaissance.

In his French commentary on Aristotle's *On the Heavens,* completed in 1377, Nicole Oresme thought the human mind was naturally inclined to conceive of spatial existence beyond our finite world. It seemed intuitively unsatisfying to believe that a finite world actually occupied all the space in existence. Extracosmic space is described as an infinite and indivisible void which is not only the immensity of God, but God Himself. By its identification with God, the extracosmic void was apparently conceived as an existent thing. Oresme even envisioned it as an infinite spatial container in which an absolute motion was at least conceivable if God chose to move our finite spherical cosmos in a straight line. To deny God's power to do this, Oresme reminds us, would be to accept article 49 (see above, p. 28) condemned at Paris one hundred years earlier. This interesting illustration of a motion deemed absolute because no other body exists outside the world to which the latter's motion could be related was also proposed by Samuel Clarke against Leibniz in the famous Clarke-Leibniz correspondence of 1715–1716. In his defense of Newton's absolute space Clarke argued that if space is but a relationship between coexistent things, as Leibniz would have it, and God moved the entire finite world, the latter could not be said to have experienced any motion, since no other thing exists outside of it to which it could bear any relationship. As would Oresme, Clarke deemed it absurd to deny the occurrence of motion under the conditions assumed.

Identification of God's immensity and infinite extension with an infinite void beyond the world raised some puzzling questions which received scant attention during the Middle Ages. Is the infinite, God-filled void a physical, three-dimensional space or a nondimensional entity? Unlike the Stoics, who, as we saw, proclaimed a three-dimensional but spirit-free infinite void, medieval Christians were confronted with a difficult problem fraught with serious theological implications. Despite vague, meager, and often ambiguous evidence, it appears that they assumed, for the most part, a dimensionless infinite void. Certain Jesuit commentators in the sixteenth century, who were intimately acquainted with earlier medieval interpretations and often reflected the opinions of their predecessors, denied that the infinite void beyond the world was a true quantity with genuine dimensions. It possessed no real or positive character, since this would have implied independent co-eternality with God. The usual description of it as "imaginary" may even derive from its alleged absence of true dimensionality and positive reality. Although it is sometimes described by terms like length, width, height, or depth, as if it were dimensional, such descriptions are to be taken in a special transcendent sense necessitated by the inadequacy of ordinary language when applied to the "dimensions" of a divine being utterly lacking in dimension as commonly understood. When, as in one case, a medieval author, Johannes de Ripa, who wrote around 1350, suggested that the infinite void is actually three-dimensional, he distinguished the dimension of infinite space from God's extensive immensity. These are not coextensive, since the latter is said to circumscribe and infinitely exceed the imaginary infinite void, which is conceived as unreal and lacking in positive attributes.

Development of the concept of an omnipresent God in an infinite void beyond the created, vacuumless world cannot be explained in isolation from the powerful intellectual currents let loose by the condemnation of 1277. The emphasis on God's absolute power, with its numerous and sometimes strange consequences, pervaded the theological, philosophical, and scientific thought of the fourteenth century. That the presence of an absolutely powerful God should extend no further than the finite cosmos of His own making simply because Aristotle had denied extramundane existence must have seemed a vexatious and unwarranted restriction on the deity. Aristotle's demonstration of this had involved the very kind of appeal that had fallen under a dark theological and philosophical cloud in the fourteenth century. The spirit of 1277 had compelled a new look at *possibilities* lying outside traditional cosmology and physics, even if it provided few new realities and laws. Article 49 of the condemnation (see above, p. 28) was found to be relevant to the problem about extracosmic void and was cited in that connection by Bradwardine and Oresme. To deny God's power to move the world in a rectilinear motion, despite the vacuum that would be left behind, invited excommunication. Reflection on the consequences of article 49 may eventually have focused attention on the succession of void places that would be left behind as God moved the world in a straight line. All of these empty places lay outside the finite cosmos which was moving into and out of them. Hence if something existed beyond the world—even if only void space—would it not be reasonable to suppose, that if God is omnipresent *inside* the world, He would also be present in whatever exists *outside* the world? But, and this is the next link in our chain of speculative reconstruction, if what lies beyond the world is to contain an infinite God, would it not be appropriate, and even necessary, that this extramundane void also be infinite? Indeed, what plausible arguments could be offered for assuming it finite? Where would it terminate? But if God is omnipresent in an infinite void beyond the world and coextensive with it, it is surely not as an extended three-dimensional entity, since God cannot be dimensional in any actually extensive or corporeal sense. Hence the imaginary infinite void was conceived as dimensionless, or dimensional in a transcendent sense, as the frequently recurring adjective "imaginary" may have sought to convey. In such a manner, perhaps, some scholastics were led to believe in the existence of a dimensionless imaginary infinite void space filled by, and associated with, an omnipresent God.

Some of the basic medieval ideas about an imaginary infinite, extra-cosmic void continued to exert influence in the seventeenth and early eighteenth centuries. Samuel Clarke, for example, insisted on God's presence in the infinite void, which he deemed void of body only. Otto von Guericke, in his justly celebrated *New Magdeburg Experiments on Void Space,* published in 1672, raised the Stoic question of what lay beyond our finite, spherical world and concluded that an infinite, imaginary void filled with God existed there. Aware that many shades of meaning had come to be attached to the term "imaginary," von Guericke thought it useful to mention some of them. Imaginary space was conceived as nothing; empty of all reality; the negation of all being; a completely fictitious entity; and God Himself, who is diffused everywhere.

If the medieval influence in all this is unmistakable, new departures are also apparent. Von Guericke identified infinite, imaginary, void space with true space, which he described as a real and positive three-dimensional entity, a description which became commonplace in the seventeenth century. Experiments and discoveries about atmospheric pressure—especially by Pascal, von Guericke, and Robert Boyle—had shown that nature did not abhor a vacuum, as was believed in the Middle Ages, and that atmospheric pressure decreased as the height of the air above the earth's surface increased. By use of pumps, artificial vacua were produced in enclosed volumes and, although there were those who denied the interpretations placed upon these experiments, others eagerly accepted the actual existence of both artificial and natural vacua. Moreover, ancient atomism, with its three-dimensional void space extending to infinity, also had its adherents, as did Stoic cosmology, in which a sealed finite world was surrounded by a real three-dimensional extramundane void. The cumulative impact of these different sources may have encouraged proponents of void to confer upon it real existence and tridimensionality. Three-dimensional vacua had become part of a new physics.

For those who also adopted the medieval view that God is omnipresent in the infinite void, there was now a genuine dilemma. If God fills a three-dimensional space, would He not have to be conceived as an actual physically extended being? But if He is nondimensional, in what sense can He be said to fill a three-dimensional space? In the Middle Ages, this dilemma was largely avoided and ignored, not only for theological reasons, but also because the problem of extracosmic, infinite void was divorced from ordinary Aristotelian physics and cosmology as these were applied to the regular operations of the world. Perhaps, among other reasons, this explains scholastic acceptance of the reality of extramundane void space—after all, God could create a body there—and rejection of its actual dimensionality in favor of transcendent meanings of terms like "extension" and "dimension" as applied to God. The situation in the seventeenth and eighteenth centuries was, however, quite different. Convinced that void space is three-dimensional and that God is omnipresent in it, many were driven to conceive of God as an extended three-dimensional being. Joseph Raphson, for example, believed that only if God was truly extended in space could He be omnipresent, since His omnipresence was a necessary prerequisite for the existence of all things. Indicating an acquaintance with medieval arguments, he declares disapprovingly that the Schoolmen had conceived of God's extension as transcendent. But how, he asks, can extended beings come from something that is only transcendently, and not actually, extended? Newton, Samuel Clarke, Henry More, and others shared Raphson's viewpoint. Not only did they associate God with a physically extended infinite void, but, on occasion, spoke as if God were actually extended in physical space, a tendency which Leibniz sought to discourage because it would transform God into a three-dimensional and corporeal being. Spinoza, indeed, took this very step when he made extension an attribute of God, who was conceived as an infinite material being. Whereas in the Middle Ages God's nondimensionality determined the properties of an imaginary and dimensionless, extramundane, infinite, void space, in the seventeenth and early eighteenth centuries it was the three-dimensional infinite void of the new physics that, unwittingly perhaps, conferred tridimensionality upon the deity.

CHAPTER VI

Conclusion

W E HAVE ELABORATED only a few of the many relevant topics that might have been included in this sketch of the salient features, and general fate, of medieval Aristotelian science. However, some tentative conclusions may be drawn from them.

Clearly, medieval science was no slavish repetition or trivial elaboration of the thoughts and opinions of Aristotle. The major figures in medieval science found much to criticise in Aristotle, not only on theological, but also on straightforward scientific grounds. Of the new theories formulated during the few hundred years of medieval scholastic science, some were developed because Aristotle's response to a given problem was deemed unsatisfactory, for example, impetus, extracosmic void, and the exponential law of motion. Some represented efforts to present equally plausible alternatives, for example, the plurality of worlds and rotation of the earth. Yet others arose in essential independence of Aristotle, for example, the intension and remission of forms and qualities, and the mean speed theorem.

Until the sixteenth and seventeenth centuries, however, such departures failed to produce genuine efforts to reconstruct, or replace, the Aristotelian world picture. The explanation for this is not easy to find. Among the factors mitigating against the overthrow or outright repudiation of the Aristotelian system, the foremost may have been its highly integrated structure. The rejection of certain crucial parts would have caused the collapse of much of the rest of it. As it happened, what was altered was frequently changed in accordance with Aristotelian principles. New changes and additions were often drawn to Aristotelian specifications and, although they were sometimes quite discordant, were again made part of the system. Thus impetus theory replaced the external contact of air with the push of an incorporeal force; lightness and heaviness, a pair of opposites basic to Aristotelian physics, were made to function as opposing forces (motive force and internal resistance) in mixed bodies; and the rectilinear motion of the earth, a notion at variance with Aristotle's opinion, was explained by continual shifts in the earth's center of gravity as it sought to coincide with the geometric center of the universe, the latter being a basic feature of Aristotelian cosmological and physical doctrine.

These numerous additions and changes were disparate, and each was formulated in response to a separate problem and tradition. They remained unrelated and sometimes inharmonious parts of the Aristotelian system rather than opening wedges that would crumble the edifice. The battered Aristotelian cosmos weathered many changes and additions over the course of a few centuries. Indeed not even the major kinematic propositions developed so brilliantly in the fourteenth century were ever assembled into a larger whole pointing toward a new mechanics. This great achievement had to await Galileo, whose success stemmed in part from a recognition of the need to disengage what was properly mathematizable from the large and diffuse context of qualitative variation.

The extraordinary tenacity of the Aristotelian system may have been strengthened and intensified by the fact that many new concepts, even those seemingly incompatible with its physics and cosmology, were enunciated in a hypothetical form and the numerous consequences derived from them were not seriously applied to nature. As we have already stressed, this state of affairs developed in the aftermath of the condemnation of 1277. In one sense, of course, the condemnation weakened the hold of Aristotelian science and philosophy on the learned world. The certainty and confidence that had characterized Aristotelian natural philosophers in the thirteenth century were undermined. Emphasis on God's absolute power, coupled with legitimate critiques of the foundations of scientific and philosophic certitude, altered considerably the character and scope of scientific discussion. Alternatives and possibilities that were undreamt of and undiscussed in the thirteenth century were raised and explored in the fourteenth. Had the consequences of the new ideas and concepts been vigorously pursued with an eye to their application to nature, medieval faith in the Aristotelian system might have been totally destroyed and something new devised to replace it. However, although confidence in the Aristotelian physical system was certainly weakened, it was largely because of a widespread lack of confidence in physical explanations generally. The great majority of scholastic authors had no desire to destroy the Aristotelian world picture and, all things considered, they probably judged it to be as reasonable and satisfying a physical system as they were apt to have. In the aftermath of the condemnation of 1277, with its emphasis on God's absolute power, their objective was to demonstrate that alternatives to a variety of Aristotelian physical explanations were not only logically possible but in some cases even as plausible as Aristotle's. The results of this approach were intellectually stimulating and productive, but were by no means such as to weaken seriously the dominant position of the Aristotelian physical system. No serious rival to it was even proposed until the sixteenth century. Even then, despite the publication of Copernicus' revolutionary heliocentric system, whose consequences might ordinarily have hastened its demise, the Reformation and Counter-Reformation served to entrench Aristotle's physics and cosmology more deeply and inflexibly than ever before. The fascinating, and sometimes significant, anti-Aristotelian alternatives so enthusiastically discussed in the fourteenth century were now ignored and often forgotten. Protestants and Catholics alike clung tenaciously to Aristotle's cosmology while vigorously denouncing Copernicus. Only in the seventeenth century did Copernicus' heliocentric system supplant the geocentric Aristotelian cosmology; and only then did the physical consequences derived from the assumed daily and annual motion of the earth destroy Aristotle's physics as well.

But why were these events so long delayed? Should they not have occurred much earlier? After all, scholastics like Buridan and Oresme had already discussed the possible diurnal rotation of the earth on its axis and anticipated some of Copernicus' arguments. Had they known of Aristarchus of Samos' full heliocentric assumption, they would have undoubtedly considered, in much the same fashion, whether the earth also moved annually around the sun, or vice versa. Perhaps the heliocentric system, which represents the first momentous step on the path to the scientific revolution, might have been proclaimed in the fourteenth century. Or perhaps some other revolutionary theory might have been enunciated. In fact, however, such an occurrence would have been most improbable. Among a number of speculative arguments that might be invoked to show the implausibility of it only one will be pursued here. It concerns attitudes toward "saving the phenomena" and claims about physical reality.

The condemnation of 1277 and the philosophical and theological consequences that flowed from it in the fourteenth century created an unusual intellectual climate in science and philosophy. No longer was it widely believed that certainty could be acquired about causes and laws of nature. It was now a matter of choosing the most probable of a number of alternatives. Even those who believed inwardly that acquisition of scientific truth was possible—usually masters of arts—were forced by the change of attitude to couch their conclusions in hypothetical language. A sophisticated positivistic attitude developed in which many of the fourteenth century Mertonians and Parisians, who contributed most to fourteenth century scientific thought, abandoned hope of acquiring true knowledge of the physical world. Instead, they channeled their energies into hypothetical discussions such as intension and remission of forms; or they conjured up fascinating hypothetical problems, as, for example, the way bodies would behave if placed in a vacuum that God had created between the moon and earth. Although, as we have seen, many interesting, significant, and often ingenious responses were made to a wide range of hypothetical questions, little was made of the results that proved directly beneficial to the advance of science. They were clearly content to exercise their scholastic ingenuity on hypothetical problems "according to the imagination" (*secundum imaginationem*). Solutions to such problems were not intended for application to nature. Logical consistency, not a quest for physical reality, was the major objective. Certain of the topics discussed in this context, and the conclusions and proposed solutions that emerged, were destined to play a vital role as the scientific revolution unfolded in the sixteenth and seventeenth centuries. Examples are the discussions on the diurnal rotation of the earth, the kinematic theorems embedded in treatises on intension and remission of forms, and the investigations concerning hypothetical vacua. But these ingenious conclusions and solutions could play no role in the development of a new science until they were divorced from an approach to nature characterized by *secundum imaginationem* and "saving the phenomena" and associated instead with a quest for physical reality. Only then could significant new paths emerge in the history of science. Not until those who proposed anti-Aristotelian, or at least non-Aristotelian, scientific theories, laws, and explanations were also moved to discover physical reality was there a chance of destroying Aristotelian science and replacing it with a new cosmology and physics. With Copernicus something like this did indeed happen.

In contrast to his fourteenth century predecessors, Copernicus revealed a radically different attitude toward science and nature. His departure from fourteenth century tradition is not to be found in the arguments he offered in support of a daily axial rotation of the earth. Indeed, many of these were commonplace in scholastic discussions. It is found, rather, in Copernicus' insistence that the earth undergoes a real physical motion and in the methodological rationale which emerged from this profound belief. For Copernicus, these two aspects were intimately related. He was convinced that the assumption of a daily and annual motion of the earth not only explained the known celestial phenomena, but revealed a simpler, and therefore more harmonious, universal order. On these grounds, he boldly proclaimed that the earth truly described an annual orbit around the sun and turned once a day on its axis. The observed daily and annual motions of the sun were mere appearances resulting from the true motions of the earth.

Asserting the reality of the earth's motions—diurnal and annual—distinguishes Copernicus from his medieval predecessors, who refused to ascribe even a single diurnal motion to the earth. The break is even more significant, however, when it is realized that *hypothesis* was one of the terms he applied to his fundamental propositions about the earth. By *hypothesis,* however, he meant neither a mere convenience to save the phenomena, nor an alternative that was more probable than others. It was a fundamental truth of the physical universe. *Only if hypotheses are true* can the appearances really be saved. The diurnal and annual motion of the earth was a hypothesis that Copernicus believed to be indubitably true. This double motion produced a symmetry in the universe that was clearly superior to the older scheme. Retrogradations and progressions of the planets were rendered physically intelligible. These consequences of the earth's motion appear to have been instrumental in convincing Copernicus that the earth really moved and that his hypotheses were a true reflection of cosmological reality.

A vast gulf separates the attitude of Copernicus from that of his fourteenth century predecessors. For Copernicus "saving the phenomena" was not a matter of convenience, but truth; for Buridan and Oresme, it was not a matter of truth, but convenience. They believed that either hypothesis could equally well save the *astronomical* phenomena. The decision in favor of the earth's immobility was made on nonastronomical grounds. For medieval speculative cosmologists it was by no means necessary that astronomical hypotheses reflect cosmological truth. Indeed, any number of different hypotheses could theoretically save physical appearances. It was even widely held that a false hypothesis might be more probable than some true ones. Copernicus, however, brought a wholly different outlook and temperament to astronomy and science. To say that two hypotheses could equally well save the astronomical phenomena would have been, for him, tantamount to a confession of ignorance and confusion. Two conflicting hypotheses on the earth's status could not, upon critical reflection, truly save the phenomena. Further criteria must be sought and, when found, would allow a proper separation of false from true. For Copernicus, to "save the phenomena" in astronomy meant to frame true hypotheses. Science and its hypotheses must treat of realities, not fictions. In this sense, if no other, Copernicus may be viewed as the first great figure of the scientific revolution. It was essentially his attitude that came to prevail with Galileo, Kepler, Descartes, and Newton.

It must not be thought that during the Middle Ages there was little interest in seeking knowledge of physical reality. On the contrary, since Aristotle himself was convinced that he had arrived at a system which represented physical reality, his many followers in the thirteenth century, most notably Thomas Aquinas, were also physical realists, much like Copernicus. Since the quest for physical reality, which was an important aspect of the scientific revolution, was also well represented by faithful and narrower Aristotelians, why was it not possible for the physical realists of the Middle Ages to generate a scientific revolution? Almost certainly the answer is that their devotion and commitment to the Aristotelian system precluded this. Their physical realism was, for the most part, indistinguishable from their wholehearted acceptance of Aristotle's physics and cosmology. Prior to the condemnation and the attitudes it generated, there would have been little likelihood of the Aristotelian physical realists producing any significant criticism of Aristotle. As devoted followers, they had little incentive to criticise him. Ironically, then, it was only when physical realism was overshadowed and largely eclipsed by a Christian positivism developed in the fourteenth century that significant criticisms of, and departures from, Aristotle's cosmology and physics were proposed and vigorously pursued. The vigor which manifested itself in the fourteenth century was but a consequence of the belief that knowledge of physical reality was virtually impossible of attainment and that alternatives to this or that Aristotelian explanation were always possible, whether plausible or not. Nominalists is a label that is frequently, if inaccurately, applied to these scholastics of the fourteenth century. They generated most of the interesting and potentially significant hypothetical results in cosmology and physics, especially kinematics, and might have destroyed the Aristotelian system if they had applied their results to physical reality. Instead, their stimulating ideas and concepts were offered as mere alternatives or imaginary solutions to hypothetical problems. What was required, but not forthcoming, was a potent union of new ideas that would challenge the traditional physics and cosmology—and here significant achievements were made—with the conviction, even if naive, that knowledge of physical reality was fully attainable.

Whatever the reason for their general scepticism about coming to know physical reality—it may have been a genuine philosophic belief, or a false posture forced upon unwilling intellects by the condemnation of 1277, or a combination of both—it was not until Copernicus that a new anti-Aristotelian cosmology was proposed, and, moreover, by one who believed he had arrived at physical truth. So profoundly did Copernicus believe this that he even subordinated physics to astronomy, thus reversing an ancient and medieval tradition. Previously, it had been the task of physics to determine the causes and fundamental principles governing the cosmos. In accordance with this knowledge, the physicist alone could explain which terrestrial and celestial bodies were best suited for motion and which for rest. By contrast, astronomy was divorced from such matters. Charting and predicting the positions of the heavenly bodies was its sole task, and astronomers were ostensibly free to employ any hypotheses and devices that might facilitate their efforts. Copernicus overturned all this by insisting, on straightforward astronomical and cosmological grounds, that his revolutionary hypotheses were true. Time-honored physical notions about the earth now had to be drastically altered. Copernicus had to devise a suitable physics which could, at the very least, take the earth's motion into account. Not surprisingly, his physics, consisting of a series of scattered remarks in Book I of his *On the Revolutions* (*De Revolutionibus*), drew upon Aristotelian and scholastic principles. The weakness of his physics, however, is far less significant than his decision, a necessity under the circumstances, to bend physics to meet the needs of his revolutionary astronomy and cosmology. Physics must adapt itself to the demands and decrees of astronomy. Thus was fashioned a momentous break with an almost sanctified tradition. Had Copernicus remained in the scholastic tradition and enunciated his hypotheses as mere conveniences and computational aids whose truth or falsity were not at issue, the great new problems in dynamics and celestial mechanics which emerged as a consequence of the earth's real motion might not have arisen. Physical doctrine would have remained in harness to an immobile earth with dynamic explanations expressed in terms of impetus theory, natural place, violent motion, and related medieval explanations.

There is no denying, however, that in the fourteenth century a high degree of philosophical sophistication was developed about the role of hypotheses in the fabric of science. Whether by compulsion or choice, few scholastic authors harbored delusions that indubitable truths about physical reality could be attained. Perhaps because of this, they produced a lively body of hypothetical science and, in some instances, anticipated significant theorems and concepts that were destined to play a large role in the new science. But a tradition that emphasized uncertainty, probability, and possibility as against certainty, exactness, and faith that fundamental physical truths—which could not be otherwise—were attainable, was not likely to produce a scientific revolution. Deep faith and belief in man's ability to acquire truth of physical reality were almost indispensable. Such a faith burned brightly and steadily within Copernicus. When Andreas Osiander, a Lutheran theologian, sought to persuade Copernicus to offer his heliocentric system as a mere astronomical device which could save the phenomena but whose hypotheses might be false, he emphatically refused. Osiander's attitude would have been commonplace in the fourteenth century. To the nominalists, who produced so much of what was intellectually stimulating in medieval natural philosophy, the Copernican system would constitute an inference transcending experience. Its admittedly greater simplicity and explanatory power would not have justified the momentous step of conferring reality upon the system. God could have made this contingent world complex rather than simple. His absolute and unpredictable power makes it virtually impossible for the human mind to acquire certain knowledge of anything other than what is immediately perceptible. By ignoring this essentially pessimistic philosophy and allowing his mind to think anew about the structure of the world, Copernicus devised a simpler cosmological model, the very simplicity of which, for him, was a guarantee of physical reality. This is the stuff of error, fantasy, and scientific revolutions.

Bibliographical Essay

The following bibliography emphasizes the topics discussed in this book, but other aspects of medieval science are also included. Byzantine and Arabic science are generally omitted. Since modern vernacular translations of original sources are obviously more relevant and useful, primary sources are cited in Latin editions in a few instances only.

General works useful for all periods and sciences in the Middle Ages are given first. Information on the lives and works of all major, and many minor, figures in the history of medieval science can be found in the *Dictionary of Scientific Biography* (New York: Scribner's Sons 1970–   ) of which the first two volumes (of 13 volumes) have already appeared. Each article is followed by a critical bibliography. Still indispensable, though slightly dated, is George Sarton, *Introduction to the History of Science* (3 vols. in 5 parts; Baltimore, 1927–48). It contains a vast amount of biographical and bibliographical information and is a truly monumental work covering every aspect of ancient and medieval science up to 1400. The *New Catholic Encyclopedia* (15 vols.; New York, 1967) includes biographical sketches as well as articles on medieval science and philosophy (the first edition, known as the *Catholic Encyclopedia,* remains useful). Begun in 1967, the *International Medieval Bibliography* (published by the Department of History, the University of Minnesota) covers all fields including science and philosophy. Titles are cited systematically from hundreds of journals. Originally issued on 3 by 5 inch index cards, the bibliographies are now issued quarterly in booklet or journal form.

ARISTOTLE

Two excellent standard English editions of the works of Aristotle are J. A. Smith and W. D. Ross (eds.), *The Works of Aristotle translated into English* (12 vols.; Oxford, 1908–1952) and the Loeb Classical Library (London and Cambridge, Mass.), where the Greek text faces the translation. A convenient one volume edition which includes the *Physica (Physics), De caelo (On the Heavens), De generatione et corruptione (On Generation and Corruption),* and *Parva Naturalia (The Short Physical Treatises),* but with only brief extracts from the biological works, is Richard McKeon (ed.), *The Basic Works of Aristotle* (New York, 1941). A stimulating and lucid introduction to Aristotle's physical and, especially, biological thought is G. E. R. Lloyd, *Aristotle: The Growth and Structure of His Thought* (Cambridge, England, 1968); less useful for its biological account, but offering a good exposition of physics and cosmology is D. J. Allan, *The Philosophy of Aristotle* (London, 1952). See also the more difficult introductory account by W. D. Ross, *Aristotle,* 5th ed. (London, 1949).

SCIENCE IN LATE ANTIQUITY AND THE EARLY MIDDLE AGES

No comprehensive histories are available for science in late antiquity or the early Middle Ages to about 1000 A.D. An excellent brief account of the fate of Greek science and the role of spiritual forces to approximately 900 A.D. is Marshall Clagett, *Greek Science in Antiquity* (New York, 1955), part II. The level and methods of education and the literature of the period are described in M. L. W. Laistner, *Thought and Letters in Western Europe A.D. 500 to 900* (rev. ed., London, 1957). Early Christian education is summarized in H. I. Marrou, *A History of Education in Antiquity* (New York, 1956), chapters 9–11. On changes in the conception of the mathematical sciences between the time of Boethius and Cassiodorus, see H. M. Klinkenberg, "Der Verfall des Quadriviums im frühen Mittelalter," in Josef Koch (ed.), *Artes Liberales; von der antiken Bildung zur Wissenschaft des Mittelalters* (Leiden and Cologne, 1959), pp. 1–32.

The interpretations and theories of Greek Stoics, Aristotelians, and Neo-Platonists on space and time, matter, sublunar mechanics, modes of physical action, celestial physics, and unity of heaven and earth are described by Samuel Sambursky, *The Physical World of Late Antiquity* (New York, 1962). Opinions of the Church Fathers and others on a wide range of natural phenomena and scientific questions are contained in commentaries on the creation of the world as described in *Genesis.* These are discussed by Pierre Duhem, *Le Système du monde* (10 vols.;

Paris, 1913–1959), Vol. 2, pp. 393–501; F. E. Robbins, *The Hexameral Literature; a Study of the Greek and Latin Commentaries on Genesis* (Chicago, 1912); and Lynn Thorndike, *A History of Magic and Experimental Science* (8 vols.; New York, 1923–1958), Vol. I.

A splendid description of the handbook tradition of Greek and Latin encyclopedists down to the period immediately preceding the translations of the twelfth century is William Stahl, *Roman Science* (Madison, Wis., 1962). The following encyclopedic handbooks are available in translation: William H. Stahl (tr.), *Macrobius, Commentary on the Dream of Scipio* (New York, 1952); Ernest Brehaut (tr.), *An Encyclopedist of the Dark Ages* (New York, 1912), which contains extracts from the twenty books of Isidore of Seville's *Etymologiae* (*Etymologies*) and includes a discussion of the seven liberal arts [medical extracts from Isidore's *Etymologiae* have been published by William D. Sharpe, M.D., "Isidore of Seville, The Medical Writings, An English Translation with an Introduction and Commentary," in *Transactions of the American Philosophical Society*, New Series, Vol. 54, part 2 (1964)]; and L. W. Jones (tr.), *Cassiodorus Senator, An Introduction to Divine and Human Readings* (New York, 1946).

THE TRANSLATION AND TRANSMISSION OF ARABIC AND GREEK SCIENCE BETWEEN THE TENTH AND THIRTEENTH CENTURIES

For the early phase of the transmission process, see J. W. Thompson, "The Introduction of Arabic Science into Lorraine in the Tenth Century," *Isis*, Vol. 12 (1929), pp. 184–193 and Mary C. Welborn, "Lotharingia as a Center of Arabic and Scientific Influence in the Eleventh Century," *Isis*, Vol. 16 (1931), pp. 188–199. For the twelfth century, see J. C. Russell, "Hereford and Arabic Science in England about 1175–1200," *Isis*, Vol. 18 (1932), pp. 14–25 and Theodore Silverstein, "Daniel of Morley, English Cosmologist and Student of Arabic Science," *Mediaeval Studies*, Vol. 10 (1948), pp. 179–196.

On the translators and translations in Spain for this period, there is a good summary account by J. M. Millas-Vallicrosa, "Translations of Oriental Scientific Works," in Guy S. Métraux and Françoise Crouzet (eds.), *The Evolution of Science* (New York, 1963), pp. 128–167. The most detailed and systematic account of translations from Arabic into Latin is Moritz Steinschneider, *Die europäischen Übersetzungen aus dem Arabischen bis Mitte des 17. Jahrhunderts* (Graz, 1956; reprint of earlier articles). The various subdivisions are arranged alphabetically for easy use. Charles H. Haskins, *Studies in the History of Medieval Science*, 2nd ed., (Cambridge, Mass., 1927) is devoted almost exclusively to translators and translations from both Arabic and Greek. A condensed version appears in Haskins' excellent volume *The Renaissance of the Twelfth Century* (New York, 1957 [1927]), chapter 9 ("The Translators from Greek and Arabic"). A. C. Crombie, *Medieval and Early Modern Science* (2 vols.; New York, 1959), Vol. I, pp. 33–64, gives a brief account but provides useful tables (pp. 37–47) citing author, work, Latin translator, and language from which the translation was made, as well as the place and date of the Latin translation. Gerard of Cremona's numerous translations from Arabic to Latin are listed by Sarton, *Introduction to the History of Science*, Vol. II, part 1, pp. 338–344; those of William of Moerbeke by Pierre Thillet, *Alexandre d'Aphrodise 'De Fato ad imperatores' version de Guillaume de Moerbeke édition critique avec introduction et index* (Paris, 1963), pp. 29–35. Lists for both translators with English titles and notes will appear in Edward Grant, *A Source Book in Medieval Science* (in press).

The science of this period was modest indeed, although it improved somewhat as the twelfth century approached and remained on an upward course through that century. Material relevant to early medieval geometry (the mathematical correspondence between Radolf of Liège and Ragimbold of Cologne, mentioned in this volume, a treatise on the quadrature of the circle by Franco of Liège, and other topics) is discussed by Paul Tannery, "La géometrie au xie siècle," *Mémoires scientifiques*, Vol. 5 (Paris, 1922), pp. 79–102; many of the same topics are also discussed elsewhere in the same volume by Paul Tannery and Abbé Clerval, "Une correspondance d' écolatres du onzième siècle," pp. 229–303. The transmission of the texts of the Roman *agrimensores*, or surveyors, who furnished the early Middle Ages with much of the little geometry they knew, is traced by B. L. Ullman, "Geometry in the Medieval Quadrivium," *Studi di bibliografia e di storia in onore di Tammaro de Marinus*, Vol. 4 (Verona, 1964), pp. 263–285. In astronomy, Pierre Duhem traces the influence of the Heraclidean astronomical system through the early Middle Ages in *Le Système du monde*, Vol. 3, pp. 44–112; see also H. Lattin, "Astronomy: Our View and Theirs," Symposium on the Tenth Century, *Medievalia et Humanistica*, fasc. IX (1955), pp. 13–17.

The emergence of the medical school at Salerno, Italy during the eleventh and twelfth centuries was a notable event in the history of medicine and marked a decided upturn in the fortunes of that discipline. An excellent account of Salerno's development is Paul O. Kristeller, "The School of Salerno," *Bulletin of the History of Medicine,* Vol. 17 (1945), pp. 138–194; reprinted in P. O. Kristeller, *Studies in Renaissance Thought and Letters* (Rome, 1956), pp. 495–551. In addition to helpful articles on Salerno ("The Rise of Medicine at Salerno in the Twelfth Century," *Annals of Medical History,* n.s., Vol. 13 [1931], pp. 1–16, and "Salernitan Surgery in the Twelfth Century," *The British Journal of Surgery,* Vol. 25 [1937–38], pp. 84–99), George W. Corner has translated four Salernitan anatomical texts in *Anatomical Texts of the Earlier Middle Ages* (Washington, D.C., 1927). For medicine in northern Europe, see Loren C. MacKinney, *Early Medieval Medicine with Special Reference to France and Chartres* (Baltimore, 1937); Wilfrid Bonser, *The Medical Background of Anglo-Saxon England* (London, 1963); and C. H. Talbot, *Medicine in Medieval England* (London, 1967). An unusual twelfth century physician and mystic in Germany is discussed by Charles Singer, "The scientific views and visions of Saint Hildegarde of Bingen," Charles Singer (ed.), *Studies in the History and Method of Science* (2 vols.; Oxford, 1917–1921; reprinted 1955), I, pp. 1–55 and Gertrude M. Engbring, "Saint Hildegard, Twelfth Century Physician," *Bull. Hist. of Medicine,* Vol. 8 (1940), pp. 770–784. On the preservation of Hippocratic medical traditions, see Loren C. MacKinney, "Medical Ethics and Etiquette in the Early Middle Ages: The Persistence of Hippocratic Ideals," *Bull. Hist. of Medicine,* Vol. 26 (1952), pp. 1–31.

During the tenth century, the teaching of science was improved by Gerbert of Aurillac, whose techniques and methods are described by O. G. Darlington, "Gerbert the Teacher," *American Historical Review,* Vol. 52 (1947), pp. 456–476. Through his students and their intellectual descendants, the school of Chartres became a famous center of learning, reaching its apex in the first half of the twelfth century. On the development and activities at the school, see J. A. Clerval, *Les écoles de Chartres au moyen âge (du Ve au XVIe siècle)* (Paris, 1895; reprinted). Many who taught, or were educated there, contributed to the growing interest in science. Adelard of Bath, for example, not only translated Euclid's *Elements* and al-Khwarizmi's astronomical tables from Arabic to Latin, but also wrote the *Questiones Naturales (Natural Questions)*, which exhibits a new intellectual confidence and concomitant scorn for the older Latin learning; a translation of it appears in Hermann Gollancz, *Dodi Venechdi (Uncle and Nephew), the Work* of *Berachya Hanakdan, now edited from the MSS. at Munich and Oxford, an English Translation, Introduction, etc. to Which is Added the First English Translation from the Latin of Adelard of Bath's Quaestiones Naturales* (London, 1920). Creation and the structure of the cosmos were intensively discussed at the school of Chartres and are treated in the following: T. Gregory, "L'idea della natura nella scuola di Chartres," *Giornale critico della filosofia italiana* (1952), pp. 433–442; M. D. Chenu, O.P., "Nature and Man at the School of Chartres in the Twelfth Century," in G. Métraux and F. Crouzet (eds.), *The Evolution of Science* (New York, 1963), pp. 220–235; J. M. Parent, *La Doctrine de la creation dans l'école de Chartres* (Paris and Ottawa, 1938); Etienne Gilson, "Platonism in the Twelfth Century," *History of Christian Philosophy in the Middle Ages* (London, 1955), pp. 139–153, and "La Cosmogonie de Bernardus Silvestris," *Archives d'histoire doctrinale et littéraire du moyen âge,* Vol. 3 (1928), pp. 5–24; and Theodore Silverstein, "The Fabulous Cosmogony of Bernardus Silvestris," *Modern Philology,* Vol. 46 (1948–1949), pp. 92–116. More occult facets of thought for Adelard of Bath, William of Conches, and Bernard Silvester are described by Lynn Thorndike, *A History of Magic and Experimental Science,* Vol. 2, chapters 36, 37, and 39.

On the changes and intellectual ferment of the twelfth century, which has been frequently called a "Renaissance," see G. Paré, A. Brunet, and P. Tremblay, *La Renaissance du XIIᵉ siècle: les écoles et l'enseignement* (Paris and Ottawa, 1933) and Charles H. Haskins, *The Renaissance of the 12th Century* (1957). The latter also contains chapters on the revival of science and philosophy in the twelfth century based upon the new translations. A qualified judgment on the propriety of designating the twelfth century as a Renaissance is given by Eva Matthews Sanford, "The Twelfth Century-Renaissance or Proto-Renaissance," *Speculum,* Vol. 26 (1951), pp. 635–642. Selections and extracts concerning the controversy over the reality of a twelfth century Renaissance and its status with respect to the Italian Renaissance appear in Charles R. Young (ed.), *The Twelfth-Century Renaissance* (New York, 1969).

UNIVERSITIES

A large body of modern literature now exists on all aspects of medieval university life. The universities of Paris, Oxford, and Bologna were prototypes of all other European universities and also great centers of scientific thought. Attention may therefore be focused on them. The most fundamental work on medieval universities is Hastings Rashdall, *The Universities of Europe in the Middle Ages* (3 vols.; new edition by F. M. Powicke and A. B. Emden, Oxford, 1936), which contains detailed bibliographies (consult them for references to other valuable histories of universities). Of the brief accounts, the most readable and exciting is Charles H. Haskins, *The Rise of the Universities* (N.Y., 1923; reprinted 1957); also interesting is Lowrie J. Daly, *The Medieval University 1200–1400* (New York, 1961). An intellectually stimulating and provocative recent volume on the universities of Oxford and Paris, which includes not only institutional and academic developments, but also doctrinal controversies (the condemnation of 1277 and the sceptical currents of the fourteenth century are considered at some length) is Gordon Leff, *Paris and Oxford Universities in the Thirteenth and Fourteenth Centuries: An Institutional and Intellectual History* (New York, 1968). Of particular relevance are Guy Beaujouan, "Motives and Opportunities for Science in the Medieval Universities," A. C. Crombie (ed.), *Scientific Change* (New York, 1963) pp. 219–236, and Vern L. Bullough, *The Development of Medicine as a Profession: The Contribution of the Mediaeval University to Modern Medicine* (New York, 1966) which includes a valuable bibliographical essay on medieval medical education. An excellent, wide-ranging collection of source readings in translation touching upon organization, statutes, curriculum, student life, controversies, and so on, is Lynn Thorndike, *University Records and Life in the Middle Ages* (New York, 1944). See also Charles H. Haskins, "Life of Mediaeval Students as Illustrated by their Letters," and "Manuals for Students" in *Studies in Mediaeval Culture* (Oxford, 1929).

On Oxford, Paris, and Bologna, the following are noteworthy.

Oxford: Charles E. Mallet, *A History of the University of Oxford,* Vol. I: *The Mediaeval University and the Colleges Founded in the Middle Ages* (New York, 1924); for biographical data on all who studied or taught at Oxford until 1500, see A. B. Emden, *A Biographical Register of the University of Oxford to A.D. 1500* (3 vols.; Oxford, 1957). Also valuable are A. G. Little, *The Grey Friars at Oxford* (Oxford, 1892) and A. G. Little and F. Pelster, *Oxford Theology and Theologians, 1282–1302* (Oxford, 1934). For the scientific treatises studied, see James A. Weisheipl, "Curriculum of the Faculty of Arts at Oxford in the Early Fourteenth Century," *Mediaeval Studies,* Vol. 26 (1964), pp. 143–185. Merton College, so significant in the physical sciences during the fourteenth century is treated by G. C. Brodrick, *Memorials of Merton College* (Oxford, 1885).

Paris: Charles Thurot, *De l'organisation de l'enseignement dans l'Université de Paris au moyen âge* (Paris and Besançon, 1850); Palémon Glorieux, *Les origines du Collège du Sorbonne* (Notre Dame, Indiana, 1959). Useful for philosophy and science is P. Féret, *La Faculté de théologie de Paris* (7 vols., Paris, 1900–1910); information on the theology masters at Paris in the thirteenth century is given by Palémon Glorieux, *Répertoire des maitres en théologie de Paris au XIIIe siècle* (2 vols.; Paris, 1933). On the study of medicine at Paris, see Pearl Kibre, "The Faculty of Medicine at Paris, Charlatanism and Unlicensed Medical Practices in the Later Middle Ages," *Bull. Hist. of Medicine,* Vol. 27 (1953), pp. 1–20; Vern L. Bullough, "The Medieval Medical University at Paris," *Bull. Hist. of Medicine,* Vol. 31 (1957), pp. 197–211, and "The Development of the Medical Guilds at Paris," *Medievalia et Humanistica,* Vol. 12 (1958), pp. 33–40.

Bologna: A general university history is A. Sorbelli, *Storia della universita di Bologna,* Vol. I: Il medio evo, saec. XI–XV (Bologna, 1940). The intellectual and social climate is described in G. Zaccagnini, "La vite dei maestri e degli scolari nello studio di Bologna nei secoli XIII e XIV," *Biblioteca dell' Archivum Romanicum,* ser. I, V (Geneva, 1926). On medicine, see Vern L. Bullough, "Mediaeval Bologna and Medical Education," *Bull. Hist. of Medicine,* Vol. 32 (1958), pp. 201–215.

Since the University of Montpellier was an important medical center in the fourteenth century, see also Sonoma Cooper, "The Medical School of Montpellier in the Fourteenth Century," *Annals of Medical History,* n.s., Vol. 2 (1930), pp. 164–195; Vern L. Bullough, "The Development of the Medical University at Montpellier to the End of the Fourteenth Century," *Bull. Hist. of Medicine,* Vol. 30 (1956), pp. 508–523; and Michael R. McVaugh, "Quantified Medical Theory and Practice at Fourteenth-Century Montpellier," *Bull. Hist. of Medicine,* Vol. 43 (1969), pp. 397–413.

THE RECEPTION OF THE WORKS OF ARISTOTLE IN THE LATIN WEST AND THEIR IMPACT ON PHILOSOPHY, SCIENCE, AND THEOLOGY

Aristotle's overwhelming and unparalleled significance for medieval science and philosophy, has generated major projects devoted to publication of the medieval (primarily Latin) translations of his works, as well as those of his important Greek and Arabic commentators. Editions of all Latin translations of Aristotle's works from Greek and Arabic will eventually be published in *Corpus philosophorum medii aevi, Aristoteles Latinus,* edited by G. Lacombe, L. Minio-Paluello, *et al.* (Bruges and Paris, 1953–   ). Among the texts published thus far are versions of Aristotle's various logical works, *Posterior Analytics, Physics, De mundo* (pseudo-Aristotle), *Politics, Poetics,* and *On the Generation of Animals (De generatione animalium).* Two separate volumes plus a supplement serve as a general introduction to the project and also list all the known medieval Latin manuscripts of Aristotle's works (*Aristoteles Latinus, Codices,* edited by G. Lacombe, E. Franceschini, L. Minio-Paluello, *et al.*: Vol. 1 [Rome, 1939]; vol. 2 [Cambridge, Eng., 1955]; and *Supplementa altera* [Bruges and Paris, 1961]). The medieval Latin translations of the extant Greek commentaries on Aristotle are in process of publication under the general editorship of G. Verbeke, *Corpus Latinum commentariorum in Aristotelem Graecorum* (Louvain and Paris, 1957–   ). Four volumes have appeared thus far (commentaries of Themistius and Philoponus on *De anima,* Ammonius on *De interpretatione,* and Alexander of Aphrodisias on the *Meteorologica*). Editions of medieval Latin and Hebrew translations of Averroes' Arabic commentaries on the works of Aristotle are being published in the series *Corpus philosophorum medii aevi, Corpus commentariorum Averrois in Aristotelem,* edited by H. A. Wolfson, D. Baneth, and F. H. Fobes under the auspices of the Mediaeval Academy of America. Though much less influential than Aristotle, editions of medieval Latin translations of Plato's works are being published under the general editorship of Raymond Klibansky, *Corpus Platonicum medii aevi, Plato Latinus* (London, 1940–   ). Plato's *Meno, Phaedo,* and the extant Latin parts of the *Parmenides* and *Timaeus* (the former accompanied by the commentary of Proclus, the latter by the commentary of Chalcidius) have appeared thus far. The series was introduced with a brief volume by Klibansky, *The Continuity of the Platonic Tradition* (London, 1939).

Further valuable information and discussion of the medieval translations of Aristotle's works can be found in Marie-Thérèse d'Alverny, "Les traductions d'Aristote et de ses commentateurs," XIIᵉ Congrès international d'Histoire des Sciences, Colloques textes des rapports, *Revue de Synthèse,* troisième série, nos. 49–52, Vol. 89 (1968), pp. 125–144; Martin Grabmann, "Methoden und Hilfsmittel des Aristotelesstudiums im Mittelalter," *Sitzungsberichte der Bayerischen*

Akademie der Wissenschaften, phil.-hist. Abteilung (Munich, 1939), Heft 5, which is an important introduction to Aristotelian translations and commentaries; S. D. Wingate, *The Mediaeval Latin Versions of the Aristotelian Scientific Corpus, With Special Reference to the Biological Works* (London, 1931; reprinted Dubuque, Iowa, n.d.); D. J. Allan, "Mediaeval Versions of Aristotle, *De caelo,* and of the Commentary of Simplicius," *Mediaeval and Renaissance Studies,* Vol. 2 (1950), pp. 82–120; and L. Minio-Paluello, "Henri Aristippe, Guillaume de Moerbeke, et les traducteurs latines médiévales des 'Meteorologiques' et du 'De generatione et corruptione' d'Aristote," *Revue philosophique du Louvain,* troisième série, Vol. 45 (1947), pp. 206–235.

Even before the translations of Aristotle's physical and biological works in the twelfth and thirteenth centuries, bits and parts of Aristotelian natural philosophy had already trickled into western Europe, a phenomenon discussed by A. Birkenmajer, "Le Rôle joué par les médecins et les naturalistes dans la réception d'Aristote au XIIᵉ et XIIIᵉ siècles," *La Pologne au VIᵉ congrès international des sciences historiques, Oslo, 1928* (Warsaw, 1930). That the source of many of these bits and parts was the popular Latin translations of the astrological works of Abu Ma'shar is argued by Richard Lemay, *Abu Ma'shar and Latin Aristotelianism in the Twelfth Century* (Beirut, 1962).

The introduction and study of Aristotle's scientific treatises is described at great length by Pierre Duhem, *Le système du monde,* Vol. 5, especially chapters 8–13, pp. 233–580; F. Van Steenberghen, *Aristotle in the West* (Louvain, 1955); and D. A. Callus, "Introduction of Aristotelian Learning to Oxford," *Proceedings of the British Academy,* Vol. 29 (1943), pp. 229–281.

The reaction to Aristotelian philosophy and science and the consequences which followed are well described in a number of histories of medieval philosophy. For two lengthy discussions, see Etienne Gilson, *History of Christian Philosophy in the Middle Ages* (London, 1955), which contains excellent bibliographical citations and is the standard work on medieval philosophy, and Frederick J. Copleston, *A History of Philosophy* (Vols. 2, 3; Westminster, Md., 1953–1957). Some of Gilson's major conclusions are adopted by David Knowles, *The Evolution of Medieval Thought* (Baltimore, 1962), who describes the main currents of medieval philosophy in an absorbing and lucid manner. Julius Weinberg, *A Short History of Medieval Philosophy* (Princeton, 1965) is a brilliant, brief, and often analytic account. In opposition to Gilson, Weinberg regards the philosophical achievements of the fourteenth century nominalists as far superior to those of their thirteenth century predecessors. Two valuable short works by Paul Vignaux, an authority on nominalism, are *Philosophy in the Middle Ages: An Introduction,* (New York, 1959) and *Le Nominalisme au XIVe siècle* (Montreal, 1948). An important article by Ernest A. Moody, "Empiricism and Metaphysics in Medieval Philosophy," *Philosophical Review,* Vol. 67 (1958), pp. 145–163, emphasizes the deep distrust and hostility of the Church and theologians toward philosophy and philosophers—a hostility that lasted until after the Protestant Reformation—and the use of philosophy by theology to discredit philosophy.

The grave apprehension of the Church and its theologians concerning the impact of Arabic and Greek science and philosophy can be gauged from the list of errors of philosophers compiled between 1270 and 1274 by Giles of Rome in his *Errores philosophorum* edited and translated by Josef Koch and John O. Riedl (Milwaukee, 1944) and by the condemnation of 219 errors in 1277 translated by L. Fortrin and Peter D. O'Neill, *Medieval Political Philosophy: A Source Book,* edited by Ralph Lerner and Muhsin Mahdi (Glencoe, New York, 1963), pp. 337–354. The most thorough discussion of the consequences of the condemnation of 1277 is Pierre Duhem, *Le Système du monde,* Vol. 6, which bears the title "Le Reflux de l'Aristotelisme: Les condemnations de 1277"; also relevant to the condemnation of 1277 is D. A. Callus, *The Condemnation of Saint Thomas at Oxford* (Oxford, 1946).

SCIENCE IN THE LATER MIDDLE AGES

Few acceptable general histories of medieval science have been written. A noteworthy exception is A. C. Crombie, *Medieval and Early Modern Science* (2 vols.; New York, 1959). A brief, reliable, and convenient outline of medieval Latin science by Guy Beaujouan appears in René Taton (ed.), *History of Science: Ancient and Medieval Science from the Beginnings to 1450,* translated from the French by A. J. Pomerans (New York, 1963), chapter 7, pp. 468–532. A good summary account of medieval intellectual currents and science (exclusive of biology and medicine) appears in E. J. Dijksterhuis, *The Mechanization of the World Picture* (Oxford, 1961), pp. 99–219 and 248–253. Although proportionately more space is devoted to Albertus Magnus and Aquinas than seems warranted, James A. Weisheipl, *The Development of Physical Theory in the Middle Ages* (London, 1959), conveys a sense of medieval scholastic science in very brief compass. Devoted almost exclusively to magic and the pseudo-sciences, but containing an enormous amount of valuable biographical and bibliographical information, is the encyclopedic work by Lynn Thorndike, *A History of Magic and Experimental Science* (8 vols.; New York, 1923–1958). The first six volumes cover the period from the first century A.D. to the sixteenth century; the final two volumes deal with the seventeenth century. See also the collection of source readings in translation by Edward Grant (ed.), *A Source Book in Medieval Science* (in press), which includes sections on physics, mathematics, astronomy, cosmology, astrology, chemistry, alchemy, geology, biology, medicine, and the condemnation of 1277.

In physical and cosmological thought, the works of Pierre Duhem mark the true beginnings of our modern knowledge and appreciation of medieval contributions to the history of science. In *Les Origines de la statique* (2 vols.; Paris, 1905–1906), Duhem revealed the existence of previously neglected and virtually unknown medieval treatises on statics. Medieval kinematics and dynamics were the subject matter of his *Études sur Léonard de Vinci* (3 vols.; Paris, 1906–1913), where he described the mean speed theorem, impetus, and other concepts. Some of these topics, and many more, were studied in his truly monumental *Le Système du monde: histoire des doctrines cosmologiques de Platon à Copernic* (10 vols.; Paris, 1913–1959). Here Duhem discusses, at considerable length, the doctrine of the possible existence of a plurality of worlds, the rotation of the earth, motion in a vacuum, and other central themes. There remains, however, a genuine need to reexamine and critically evaluate Duhem's many claims and interpretations. His obvious bias in favor of medieval science lead him to exaggerate its achievements at the expense of Galileo and seventeenth century science. Rarely did Duhem examine Latin manuscripts outside the confines of Paris, a highly risky procedure in medieval studies. In *Études Galiléennes* (3 fasc.; Paris, 1939), an epochal study of Galileo's scientific thought, Alexandre Koyré denied Duhem's claim that fourteenth century scientific achievements had influenced Galileo and played a role in generating the scientific revolution. In the postwar period, students of medieval science sought a more balanced and objective judgment of medieval achievements. The most significant response came from Anneliese Maier in a series of five remarkable volumes under the general title *Studien zur Naturphilosophie der Spätscholastik* (Rome, 1949–1958): Vol. I. *Die Vorläufer Galileis im 14. Jahrhundert* (1949); Vol.

II. *Zwei Grundprobleme der scholastischen Naturphilosophie* (2d ed., 1951), containing two monographs, Intension and Remission of Forms and Impetus Theory; Vol. III. *An der Grenze von Scholastik und Naturwissenschaft* (2d ed., 1952), covering the structure of matter, the cause of fall, acceleration, free fall in a vacuum, as well as further discussion of intension and remission of forms; Vol. IV. *Metaphysische Hintergründe der spätscholastischen Naturphilosophie* (1955); and Vol. V. *Zwischen Philosophie und Mechanik* (1958). A collection of brief articles on fourteenth century thought, only some of which are directly relevant to science, was subsequently published by Maier under the title *Ausgehendes Mittelalter* (2 vols.; Rome, 1964–1967). Crucial for all the disputes and controversies is the publication of properly edited medieval scientific texts, ideally with accompanying translations to make them available to as wide an audience as possible. Although Latin scientific texts have been published sporadically since the nineteenth century, it was Marshall Clagett who organized (1952) the series *Publications in Medieval Science* (published by the University of Wisconsin Press) and became its general editor. The texts and translations in this invaluable series have deepened and vastly extended our knowledge of medieval physics, cosmology, mathematics, and medicine. Many of the following titles appeared in this series.

CLASSIFICATION OF THE SCIENCES

The hierarchical relation between all the sciences within the general domain of knowledge and philosophy was a subject of considerable interest and significance in the Middle Ages. An excellent survey from antiquity to the Middle Ages is James A. Weisheipl, O.P., "Classification of the Sciences in Medieval Thought," *Mediaeval Studies,* Vol. 28 (1966), pp. 54–90. A briefer account restricted to the Middle Ages appears in the first part of Marshall Clagett, "Some General Aspects of Physics in the Middle Ages," *Isis,* Vol. 39 (1948), pp. 29–44; see also Joseph Marietan, *Problème de la classification des sciences d'Aristote à St. Thomas* (Paris, 1901). For two source discussions from the twelfth century, see Jerome Taylor (tr.), *The Didascalicon of Hugh of St. Victor: A Medieval Guide to the Arts* (New York, 1961), pp. 62–64 and 67–73 (Hugh reflects the Latin tradition prior to the translations of Arabic and Aristotelian science), and Marshall Clagett and Edward Grant in Edward Grant, *A Source Book in Medieval Science* (in press) where a selection from Domingo Gundisalvo's *On the Division of Philosophy* (*De divisione philosophiae*) is translated; Domingo's classification scheme was derived from Arabic sources.

PHYSICS

A fundamental work including translations and detailed analytic commentary on medieval physics is Marshall Clagett, *The Science of Mechanics in the Middle Ages* (Madison, Wis., Publications in Medieval Science, 1959). Most of the significant medieval contributions to statics, kinematics, and dynamics are described and discussed, including the mean speed theorem, impetus theory, free fall of bodies, possible rotation of the earth, and Bradwardine's law of motion. Latin texts and translations of the major medieval treatises on statics, associated largely with the name Jordanus de Nemore, have been published by Ernest A. Moody and Marshall Clagett, *The Medieval Science of Weights* (Madison, Wis., Publications in Medieval Science, 1952). The medieval statical tradition is pursued further by Joseph E. Brown, *The Scientia de ponderibus in the Later Middle Ages* (Doctoral dissertation, University of Wisconsin, 1967).

Ernest A. Moody has written a series of brilliant articles on the laws of motion and their impact on Galileo. In his most significant article, "Galileo and Avempace: The Dynamics of the Leaning Tower Experiment," *Journal of the History of Ideas*, Vol. 12 (1951), pp. 163–193 and 375–422, Moody argues that Galileo's early dynamics (the Pisan period) was derived ultimately from Avempace's law of velocities. In "Galileo and His Precursors," Carlo L. Golino (ed.), *Galileo Reappraised* (Berkeley, 1966), pp. 23–43, Moody shows that during his subsequent Paduan period, Galileo adopted Buridan's idea of a permanent impetus and his explanation of free fall, as well as Mertonian analyses of accelerated motion. Ockham's acceptance of Avempace's (and Aquinas') kinematics and rejection of its dynamic implications are discussed by Moody in "Ockham and Aegidius of Rome," *Franciscan Studies*, Vol. 9 (1949), pp. 417–442.

On the possibility of motion in a vacuum and the consequences following therefrom, see Edward Grant, "Motion in the Void and the Principle of Inertia in the Middle Ages," *Isis*, Vol. 55 (1964), pp. 265–292, and "Bradwardine and Galileo: Equality of Velocities in the Void," *Archive for History of Exact Sciences*, Vol. 2 (1965), pp. 344–364; also A. Maier, "Die freie Fall im Vakuum," *An der Grenze von*

Scholastik und Naturwissenschaft, pp. 219–254. Four medieval arguments against the actual existence of vacuum are described by Charles B. Schmitt, "Experimental Evidence for and Against a Void: The Sixteenth Century Arguments," *Isis,* Vol. 58 (1967), pp. 352–366.

The major treatise in which Thomas Bradwardine formulated his new and influential law of motion has been edited and translated by H. Lamar Crosby, Jr., *Thomas of Bradwardine His Tractatus de proportionibus Its Significance for the Development of Mathematical Physics* (Madison, Wis., Publications in Medieval Science, 1955). For Oresme's extension of Bradwardine's law to embrace irrational mathematical relations and his application of the mathematical results, see Edward Grant (ed. and tr.), *Nicole Oresme De proportionibus proportionum and Ad pauca respicientes* (Madison, Wis., Publications in Medieval Science, 1966). See also Ernest A. Moody, "Laws of Motion in Medieval Physics," *The Scientific Monthly,* Vol. 72 (1951), pp. 18–23. These and other topics are lucidly presented by Marshall Clagett, "Some Novel Trends in the Science of the Fourteenth Century," Charles S. Singleton (ed.), *Art, Science, and History in the Renaissance* (Baltimore, 1968), pp. 275–303. For analysis of a medieval treatise in which physics, mathematics, and logic are employed in the solution of sophisms and where the physical problems considered are hypothetical (*secundum imaginationem*), see Curtis Wilson, *William Heytesbury: Medieval Logic and the Rise of Mathematical Physics* (Madison, Wis., Publications in Medieval Science, 1956). Two other books on Oresme deserve mention: Marshall Clagett, *Nicole Oresme and the Medieval Geometry of Qualities and Motions, A Treatise on the Uniformity and Difformity of Intensities known as Tractatus de configurationibus qualitatum et motuum,* edited with an Introduction, English Translation and Commentary (Madison, Wis., Publications in Medieval Science, 1968), which deals magisterially with the intension and remission of forms; and Edward Grant, *Nicole Oresme and the Kinematics of Circular Motion, Tractatus de commensurabilitate vel incommensurabilitate motuum celi,* edited with an Introduction, English Translation, and Commentary (Madison, Wis., Publications in Medieval Science, 1971). The *De motu* of Galileo, frequently mentioned in this volume, has been translated in *Galileo Galilei 'On Motion' and 'On Mechanics'* comprising *De motu* (*ca.* 1590) translated with Introduction and Notes by I. E. Drabkin; and *Le Meccaniche* (*ca.* 1600) translated with Introduction and Notes by Stillman Drake (Madison, Wis., Publications in Medieval Science, 1960).

No suitable history of medieval optics has yet been written. Vasco Ronchi's *Histoire de la lumière* (Paris, 1956), a survey of the history of optics, is inadequate for the Middle Ages. An extensive, representative selection of source readings covering the whole range of medieval optics has been prepared by David C. Lindberg and will appear in Edward Grant (ed.), *A Source Book in Medieval Science* (in press). Good scholarly studies are available on special topics and aspects of medieval optics. A. C. Crombie, *Robert Grosseteste and the Origins of Experimental Science, 1100–1700* (Oxford, 1953), is concerned with scientific methodology as exemplified in medieval optics (it includes an excellent bibliography). For a Latin edition with facing translation, introduction, and critical notes of the most popular medieval optical treatise, see David C. Lindberg (ed. and tr.), *John Pecham and the Science of Optics: "Perspectiva Communis"* (Madison, Wis., Publications in Medieval Science, 1970). A study of the methodology employed in the optical treatises of Theodoric of Freiberg, who presented the first proper qualitative explanation of the primary and secondary rainbows, appears in William A. Wallace, O.P., *The Scientific Methodology of Theodoric of Freiberg* (Fribourg, Switzerland, 1959). In "Alhazen's Theory of Vision and its Reception in the West," *Isis,* Vol. 58 (1968), pp. 321–341, David C. Lindberg shows that Bacon, Pecham, and Witelo faithfully transmitted the essential and substantive contributions of Alhazen to Kepler and the seventeenth century. On the impact of philosophy and psychology on medieval optics, see Graziella F. Vescovini, *Studi sulla prospettiva medievale* (Turin, 1965).

One of the greatest empirical treatises of medieval science is *Letter on the Magnet* by Petrus Peregrinus (Pierre de Maricourt). This first systematic description of the properties of the magnet, written in the thirteenth century, has been translated into English at least three times. Of these, see Brother Arnold [i.e., Joseph Charles Mertens], *The Letter of Petrus Peregrinus On The Magnet, A.D. 1269,* with introductory notice by Brother Potamian (New York, 1904). The treatise is discussed by S. P. Thompson, "Petrus Peregrinus de Maricourt and his Epistola de Magnete," *Proceedings of the British Academy,* Vol. 2 (1905–1906), pp. 377–408.

ASTRONOMY, ASTROLOGY, AND COSMOLOGY

An authoritative summary of medieval astronomy in the Latin West has yet to be written. One may consult J. P. J. Delambre, *Histoire de l'astronomie au moyen âge* (Paris, 1819). In many parts of his *Le Système du monde,* Duhem discusses astronomical topics and E. J.

Dijksterhuis is informative, but severely limited in scope, in *Mechaniza-tion of the World Picture*, pp. 209–219. A summary and critique of the first five volumes of Duhem's *Le Système du monde* is J. L. E. Dreyer, "Medieval Astronomy," in Charles Singer (ed.), *Studies in the History and Method of Science*, Vol. 2 (Oxford, 1921), pp. 102–120. The two most popular medieval astronomical treatises were the *De spera* (*On the Sphere*) of John of Sacrobosco and the more technical *Theorica planetarum* (*Theory of the Planets*), falsely ascribed to Gerard of Cremona. The first of these has been edited and translated by Lynn Thorndike, *The Sphere of Sacrobosco and Its Commentators* (Chicago, 1949) and the second has been translated by Olaf Pedersen and will appear in my forthcoming *Source Book in Medieval Science*. A Middle English astronomical text (perhaps, in part, by Chaucer) for calculating the positions of the planets by means of a special instrument has been edited and translated in *The Equatorie of the Planetis* by Derek J. Price (Cambridge, 1955). In chapter 7, Price explains the main elements of the Ptolemaic system, on which technical medieval astronomy was based. Two treatises on comets by Albertus Magnus and Thomas Aquinas are translated in Lynn Thorndike, *Latin Treatises on Comets Between 1238 and 1368 A.D.* (Chicago, 1950). The concept of "saving the phenomena" in astronomy is traced by Pierre Duhem, *To Save the Phenomena, An Essay on the Idea of Physical Theory from Plato to Galileo*, translated from the French by Edmund Doland and Chaninah Maschler; Introductory Essay by Stanley L. Jaki (Chicago, 1969).

An interesting general sketch of medieval astrology which, however, lacks substantive discussion of astrology as a discipline, is Theodore Otto Wedel, *The Mediaeval Attitude Toward Astrology, Particularly in England* (New Haven, 1920). To my knowledge, no comprehensive Latin astrological treatise has been completely translated or analysed. Fortunately, a splendid technical Arabic treatise is available in *The Book of Instruction in the Elements of the Art of Astrology by al-Biruni*, translated by R. Ramsay Wright (London, 1934). The principles described by al-Biruni would differ little, if at all, from technical Latin astrology. Attacks by Oresme on astrological prediction can be found in G. W. Coopland (ed. and tr.), *Nicole Oresme and the Astrologers: A Study of His Le Livre de Divinacions* (Cambridge, Mass., 1952).

MATHEMATICS

Despite a reasonably large body of specialized literature, medieval mathematics has not been adequately surveyed as a whole, nor, for that matter, have its separate branches. Still the most extensive treatment is

Moritz Cantor, *Vorlesungen über Geschichte der Mathematik* (4 vols.; Leipzig, 1880–1908), Vols. 1 (3rd ed., 1907) and 2 (2nd ed., 1899–1900). A good recent survey, which considers medieval Chinese, Indian, Islamic, and European mathematics and includes a lengthy bibliography of more than 240 titles is A. P. Juschkewitsch, *Geschichte der Mathematik im Mittelalter*, (Leipzig, 1964). Selections from arithmetic, algebra, number theory, probability, infinite series, proportions, geometry, and trigonometry are included in Edward Grant (ed.), *A Source Book in Medieval Science* (in press). Although largely restricted to Archimedes and his impact on the Middle Ages, the best study of medieval theoretical GEOMETRY is Marshall Clagett, *Archimedes in the Middle Ages*, Vol. I: *The Arabo-Latin Tradition* (Madison, Wis., Publications in Medieval Science, 1964). On the fate of Euclid's *Elements*, see John E. Murdoch, "The Medieval Character of the Medieval Euclid: Salient Aspects of the Translations of the *Elements* by Adelard of Bath and Campanus of Novara," XIIe Congrès international d'Histoire des Sciences, *Revue de Synthèse*, (1968), Vol. 89, pp. 67–94. On PROPORTION and proportionality theory, see John E. Murdoch, "The Medieval Language of Proportions: Elements of the Interaction with Greek Foundations and the Development of New Mathematical Techniques," A. C. Crombie (ed.), *Scientific Change* (New York, 1963), pp. 237–271 (see also the commentaries by L. Minio-Paluello and H. Lamar Crosby, Jr., and Murdoch's response). Two basic fourteenth century treatises on proportions by Bradwardine (edited and translated by H. Lamar Crosby, Jr.) and Oresme (edited and translated by Edward Grant) have been cited earlier under laws of motion in the section on Physics. In the *De proportionibus proportionum*, Oresme arrived at the concept of an irrational exponent. To see how he manipulated rational exponents, see Edward Grant (tr.), "Part I of Nicole Oresme's *Algorismus proportionum*," *Isis*, Vol. 56 (1965), pp. 327–341. On ARITHMETIC, see Robert S. Steele (ed.), *The Earliest Arithmetics in English*, Early English Text Society, extra series, no. 118 (London, 1922). An English translation of a French arithmetic is given by E. G. R. Waters, "A Fifteenth Century French Algorism from Liège," *Isis*, Vol. 12 (1929), pp. 194–236; see also Dorothy V. Schrader, "The Arithmetic of the Medieval Universities," *Mathematics Teacher*, Vol. 60 (1967), pp. 264–278. In ALGEBRA, two works by Arabic authors formed the starting point for medieval algebra in the West. Both are available in English translation: Louis Chester Karpinski and John Garrett Winter, *Contributions to the History of Science, Part I: Robert of Chester's Latin Translation of the Algebra of al-Khowarizmi,* University of Michigan Studies, Humanistic Series, Vol. 11 (Ann Arbor, 1930) and *The Algebra of Abu-Kamil,* Hebrew Text, Translation and Commentary by Martin Levey (Madison, Wis., Publications in Medieval Science, 1966). The *De numeris datis* (*On Given Numbers*), an algebraic treatise by Jordanus de Nemore, one of the best of medieval mathematicians, has been edited by Maximilian Curtze with modern symbolic analytic summaries following each proposition; see "Commentar zu dem 'Tractatus de Numeris Datis' des Jordanus Nemorarius," Historisch-literarische Abtheilung der *Zeitschrift für Mathematik und Physik* (Leipzig, 1891), Vol. 36, pp. 1–23, 41–63, 81–95, and 121–138. Leonardo Pisano's *Liber Quadratorum* (*Book of Square Numbers*), written in 1225 and concerned with NUMBER THEORY and indeterminate analysis, has been translated into French by Paul Ver Ecke, *Léonard de Pise, Livre des nombres carrés* (Bruges, 1952). On TRIGONOMETRY, see John David Bond, "The Development of Trigonometric Methods Down to the Close of the Fifteenth Century," *Isis,* Vol. 4 (1921–1922), pp. 295–323. Bond also translated a fourteenth century trigonometric treatise in "Richard Wallingford's Quadripartitum (English Translation)," *Isis,* Vol. 5 (1923), pp. 339–358. Wallingford belonged to the distinguished group of Oxford mathematicians and physicists of the first half of the fourteenth century.

CHEMISTRY AND ALCHEMY

Because of their obscurity, relatively few medieval alchemical texts have been edited and analyzed. Pioneer work, however, has been done by Pierre E. M. Berthelot in a series of volumes, among which are *Introduction à l'étude de la chimie des anciens et du moyen âge* (Paris, 1889) and *Histoire des sciences: la chimie au moyen âge* (3 vols., Paris, 1893). The most useful and instructive single-volume work is E. J. Holmyard, *Alchemy* (Harmondsworth, Middlesex and Baltimore, 1957), which includes a glossary of terms; see also F. Sherwood Taylor, *The Alchemists, Founders of Modern Chemistry* (New York, 1949). On apparatus, see E. J. Holmyard et al. (eds.), *A History of Technology,* Vol. 2 (Oxford, 1956), pp. 731–752. For a medieval description of chemical operations, procedures, and materials presented in a straightforward, nonoccult manner, see *Libellus de alchimia ascribed to Albertus Magnus,* translated from the Borgnet Latin edition, introduction and notes by Sister Virginia Heines, S.C.N., with a foreword by Pearl

Kibre (Berkeley and Los Angeles, 1958). A scholastic defense of the alchemical art and its ability to convert base metals to gold by Petrus Bonus (*ca.* 1330) has been translated by A. E. Waite, *The New Pearl of Great Price. A Treatise Concerning the Treasure and Most Precious Stone of the Philosophers. On the Method and Procedure of this Divine Art; . . .* (London, 1894). See also Vincent R. Larkin, "Saint Thomas Aquinas: 'On the Combining of the Elements,'" *Isis,* Vol. 51 (1960), pp. 68–72. For an occult and mystical approach in which it is difficult to determine whether we have before us a religious treatise expressed in the language of alchemy or an alchemical treatise cloaked in the language of religious mysticism, see *Aurora Consurgens, A Document Attributed to Thomas Aquinas on the Problem of Opposites in Alchemy,* edited with a commentary by Marie-Louise von Franz; a companion work to C. G. Jung's *Mysterium Coniunctionis* translated by R. F. C. Hull and A. S. B. Glover (New York, 1966), Bollingen Series, Vol. 77.

GEOLOGY AND GEOGRAPHY

In *Le Système du monde,* Duhem devotes much of Vol. 9 to geology. An excellent work is *Albertus Magnus Book of Minerals,* translated by Dorothy Wyckoff (Oxford, 1967). On an experiment by Albertus Magnus to determine if volcanoes are caused by subterranean steam pressure, see Rolf A. Koch, "Die aktualistische Bedeutung der Vulkanexperimente des Albertus Magnus," *Abhandlungen des Staatlichen Museums für Mineralogie und Geologie zu Dresden,* Vol. 11 (1966), pp. 307–314. In a work that was available in Latin translation, Avicenna discussed mountain formation; see *Avicennae 'De congelatione et conglutinatione lapidum' being sections of the Kitab al-Shifa."* The Latin and Arabic Texts edited with an English translation of the latter and with critical notes by E. J. Holmyard and D. C. Mandeville (Paris, 1927). Two useful and interesting books on geography are: John K. Wright, *The Geographical Lore of the Time of the Crusades* (New York, 1925) and George H. T. Kimble, *Geography in the Middle Ages* (London, 1938). Traditional medieval geographical knowledge is contained in Pierre d'Ailly's *Ymago Mundi,* which has been published with Latin text and facing French translation by Edmond Buron, *Ymago Mundi de Pierre d'Ailly* (3 vols.; Paris, 1930), Vol. I.

TECHNOLOGY

In a field where written records by practitioners and innovators are scarce, Lynn White, Jr., *Medieval Technology and Change* (Oxford,

1962), describes brilliantly the impact on society of the stirrup, the plough, horsepower, and mechanical sources of power. Two stimulating articles by the same author are "What Accelerated Technological Progress in the Western Middle Ages?" in A. C. Crombie (ed.), *Scientific Change* (New York, 1963), pp. 272–291 (see also the discussions on pp. 311–314 and 327–331), and "Medieval Uses of Air," *Scientific American*, Vol. 223, No. 2 (August, 1970), pp. 92–100. In the latter, White explains that, prior to the mid-nineteenth century, science and technology were largely separate entities. Besides the blast furnace, windmill, and primitive suction pump, he shows that medieval workers also constructed a manned glider, gas turbine, and conceived the parachute. An extremely valuable brief account is Bertrand Gille, "Technological Developments in Europe: 1100–1400," Guy S. Métraux and François Crouzet, (eds.), *The Evolution of Science* (New York, 1963), pp. 168–219; see also R. J. Forbes, "Metallurgy and Technology in the Middle Ages," *Centaurus*, Vol. 3 (1953), pp. 49–57; reprinted in Robert Palter (ed.), *Toward Modern Science* (New York, 1969), pp. 257–267. Specialized articles on medieval mining, metallurgy, manufacture, transport, practical mechanics, and chemistry are available in Vol. 2 of Charles Singer, E. J. Holmyard *et al.* (eds.), *A History of Technology* (5 vols.; New York and London, 1954–1958). An extraordinary and complicated astronomical clockwork designed and constructed in the fourteenth century is described by Silvio A. Bedini and Francis R. Maddison, "Mechanical Universe, the Astrarium of Giovanni de' Dondi," *Transactions of the American Philosophical Society*, new series, Vol. 56, part 5 (Philadelphia, 1966).

BIOLOGY

The standard histories of biology such as Erik Nordenskiöld, *The History of Biology* (New York, 1929) and Charles Singer, *A History of Biology to About the Year 1900* (rev. ed.; New York, 1959) are worthless for the Middle Ages. Perhaps the best treatment of medieval biology is H. Balss, *Albertus Magnus als Biologe* (Stuttgart, 1947). On Albertus' knowledge of ornithology, see S. Killermann, *Die Vogelkunde des Albertus Magnus 1270–80* (Regensburg, 1910). Medieval embryology, including an appraisal of Albertus Magnus, is summarized briefly in Joseph Needham, *A History of Embryology* (2nd ed.; New York, 1959), pp. 86–96. A translation from Albertus Magnus's *Questions on Aristotle's De animalibus* will appear in Edward Grant (ed.), *A Source Book in Medieval Science*. In the same volume, I have also translated a few of the descriptions of animals which Albertus made in his extensive zoological treatise, *Twenty-Six Books on Animals* (*De animalibus libri XXVI*). These should be compared to the manifestly inferior descriptions found in a bestiary translated by T. H. White, *The Book of Beasts being a translation from a Latin Bestiary of the Twelfth Century* (London, 1954). The latter is an extension of the famous *Physiologus,* a moralizing bestiary written originally in Greek in late antiquity. During the next one thousand years it was translated into an incredible number of languages, including Latin, and thus became a truly worldwide book. The most extraordinary zoological treatise of the Middle Ages, one greatly admired by modern zoologists, is the *De arte venandi cum avibus* (*On the Art of Hunting with Birds*), written around 1245 by Frederick II, Holy Roman Emperor. A detailed study of the habits and training of falcons based upon direct observation, it has been translated by Casey A. Wood and F. Marjorie Fyfe, *The Art of Falconry being the 'De arte venandi cum avibus' of Frederick II of Hohenstaufen* (Stanford University Press, 1943).

In BOTANY, the fundamental study is Hermann Fischer, *Mittelalterliche Pflanzenkunde* (Munich, 1929). Also relevant are Agnes Arber, *Herbals: Their Origin and Evolution 1470–1670* (new ed., Cambridge, 1938), and Eleanor S. Rohde, *The Old English Herbals* (New York, 1922). In botany, as in zoology, Albertus Magnus emerges as the major figure. His *De vegetabilibus* (*On Plants*) includes theoretical and descriptive parts, with more of the former than the latter. I have translated representative selections from the theoretical and philosophical part in Edward Grant (ed.), *A Source Book in Medieval Science.* A description of the oak tree, demonstrating Albertus' observational powers, has been translated by Charles Singer, in "Greek Biology and its relation to the Rise of Modern Biology," *Studies in the History and Method of Science,* Charles Singer (ed.), Vol. 2, pp. 74–145. Definitions of botanical terms employed by Albertus Magnus are listed in two articles by T. A. Sprague in *Bulletin of Miscellaneous Information,* Royal Botanic Gardens, Kew (1933): "Plant Morphology in Albertus Magnus," pp. 431–432; and "Botanical Terms in Albertus Magnus," pp. 440–459.

MEDICINE

The literature on medieval medicine is extensive. Most general histories of medicine include a section, usually inadequate, on medieval medicine. Standard histories of medicine and its various branches are listed by George Sarton, *Horus: A Guide to the History of Science* (Waltham,

Mass., 1952), pp. 184–190. Specific histories of medieval medicine are: E. Riesman, *The Story of Medicine in the Middle Ages* (New York, 1935); J. J. Walsh, *Medieval Medicine* (London, 1920); and, despite the title, C. H. Talbot, *Medicine in Medieval England* (London, 1967) has much that is useful and authoritative on medicine in Western Europe and excels the two preceding volumes. A fascinating pictorial volume with interesting commentary is Loren C. MacKinney, *Medical Illustrations in Medieval Manuscripts* (London, 1965), which includes 18 color plates and 86 illustrations. An extensive and excellent representative selection embracing the whole range of medieval medicine has been organized by Michael R. McVaugh and will appear in Edward Grant (ed.), *A Source Book in Medieval Science.*

No complete translation has yet been made of Avicenna's *Canon of medicine,* the most comprehensive and widely used medical textbook of the Middle Ages. Two translations of Book I (there are five books in all) have appeared: a somewhat loose and inaccurate translation by O. Cameron Gruner, *A Treatise on the Canon of Medicine of Avicenna* (London, 1930), and a more recent translation by Mazhar H. Shah, *The General Principles of Avicenna's Canon of Medicine* (Karachi, 1966), which includes the section on anatomy omitted by Gruner.

On ANATOMY, see Charles Singer, *A Short History of Anatomy from the Greeks to Harvey* (New York, 1957), pp. 62–89. A translation of the most important medieval textbook on anatomy by Mondino de'Luzzi is given in Charles Singer (tr.), *The Fasciculo di Medicina, Venice, 1493, with an Introduction by Charles Singer* (Florence, 1925), Vol. I. Also valuable for medieval anatomy are *Iacopo Berengario Carpi, A Short Introduction to Anatomy (Isagogae Breves)* translated with an introduction and notes by L. R. Lind (Chicago, 1959), and Charles Singer, "A Study in Early Renaissance Anatomy, with a new text: The *Anathomia* of Hieronymo Manfredi, transcribed and translated by A. Mildred Westland," in *Studies in the History and Method of Science,* Charles Singer (ed.), Vol. I, pp. 79–164. See also the anatomical texts translated by G. W. Corner and cited earlier.

A history of medieval SURGERY, containing 162 plates and an extensive bibliography, is Pierre Alphonse Huard, and Mirko Drazen Grmek, *Mille ans de chirurgie en occident, Vᵉ–XVᵉ siècles* (Paris, 1966). At least three major surgical texts have been translated: *The Surgery of Theodoric ca. A.D. 1267,* translated from the Latin by Eldridge Campbell and James Colton (2 vols.; New York, 1955); *Lanfranc's Science of Cirurgie,* Early English Text Society, 1894, original series 102; and

Maître Henri de Mondeville, Chirurgie, traduction francaise avec des notes, une introduction et une biographie by E. Nicaise (Paris, 1893). A selection from Guy de Chauliac's medieval surgical work, which describes attitudes toward surgery and the prerequisites of a good surgeon, has been translated by James Bruce Ross, *The Portable Medieval Reader,* edited by James Bruce Ross and Mary Martin McLaughlin (New York, 1949), pp. 640–649. On dissection, see Mary Niven Alston, "The Attitude of the Church towards Dissection before 1500," *Bulletin of the History of Medicine,* Vol. 8 (1940), pp. 221–238. Hospitals are discussed in Rotha M. Clay, *The Medieval Hospitals of England* (London, 1909), and C. A. Mercier, *Leper Houses and Mediaeval Hospitals* (London, 1915). On the black death, see Philip Ziegler, *The Black Death* (New York, 1969), a lengthy description in which the emphasis is on England; it contains an extensive bibliography. An older but still interesting account with illustrative plates and bibliography is Johannes Nohl, *The Black Death: A Chronicle of the Plague* (London, 1926). An analysis of sixteen plague tractates written between 1348–1350 appears in Anna Montgomery Campbell, *The Black Death and Men of Learning* (New York, 1931). A speculative analysis of the impact of the black death and what, if anything, can be learned from it in the event of a modern disaster of similar magnitude (for example, thermonuclear war) is presented by Jack Hirshleifer, *Disaster and Recovery: The Black Death in Western Europe* (Santa Monica, Calif., 1966).

THE RELEVANCE OF MEDIEVAL PHYSICAL SCIENCE TO THE SCIENTIFIC REVOLUTION

An essential discontinuity between medieval physical science and the achievements of Galileo and the scientific revolution of the seventeenth century is maintained by Alexandre Koyré in *Études Galiléennes* (3 fascicules; Paris, 1939), "Galileo and Plato," *Journal of the History of Ideas,* Vol. 4 (1943), pp. 400–428 (reprinted in Philip P. Wiener and Aaron Noland, *Roots of Scientific Thought* [New York, 1957], pp. 147–175), and "Les origines de la science moderne," *Diogène,* Vol. 26 (1956), pp. 14–42. Ernan McMullin has also opted for discontinuity in "Medieval and Modern Science: Continuity or Discontinuity?" *International Philosophical Quarterly,* Vol. 5 (1965), pp. 103–129 and "Empiricism and the Scientific Revolution," in Charles S. Singleton (ed.), *Art, Science, and History in the Renaissance* (Baltimore, 1968), pp. 331–369. Edward Rosen has argued that modern scholarship has upheld Burckhardt's judgment that modern science began in the Renaissance when medieval scholastic attitudes toward nature were largely abandoned; see Rosen's article "Renaissance Science as Seen by Burckhardt and his Successors," in Tinsley Helton (ed.), *The Renaissance, A Reconstruction of the Theories and Interpretations of the Age* (Madison, Wis., 1961), pp. 77–103. On the side of scientific continuity and arguing for a direct medieval influence on the scientific achievements of the scientific revolution is Pierre Duhem in many of the works cited above. Two authors who have maintained that seventeenth century science inherited a well-formulated and essential methodology from the Middle Ages are John Herman Randall, Jr., and Alistair C. Crombie. For John Herman Randall, Jr., see "The Development of Scientific Method in the School of Padua," *Journal of the History of Ideas,* Vol. 1 (1940), pp. 177–206, partially reprinted in Wiener and Noland (eds.), *The Roots of Scientific Thought,* pp. 139–146, and *The School of Padua and the Emergence of Modern Science* (Padua, 1961); for A. C. Crombie, see *Robert Grosseteste and the Origins of Experimental Science 1100–1700* (Oxford, 1953), chapter 11, pp. 290–319, and "The Significance of Medieval Discussions of Scientific Method for the Scientific Revolution," in Marshall Clagett (ed.), *Critical Problems in the History of Science* (Madison, Wis., 1959), pp. 79–101 (in the same volume, see also the comments on Crombie's paper by I. E. Drabkin, pp. 142–147, and by Ernest Nagel, pp. 153–154). A brilliant middle-of-the-road view, which credits medieval scholastic science with notable achievements but takes into account the deficiencies and weaknesses that prevented a major breakthrough before Galileo, is presented by Anneliese Maier in *Ausgehendes Mittelalter* (Rome, 1964), chapters 16 ("Die Stellung der scholastischen Naturphilosophie in der Geschichte der Physik," pp. 413–424) and 17 ("Ergebnisse der spätscholastischen Naturphilosophie," pp. 425–457). Also relevant to this problem are Edward Grant, "Late Medieval Thought, Copernicus, and the Scientific Revolution," *Journal of the History of Ideas,* Vol. 23 (1962), pp. 197–220 and Ernest A. Moody, "Galileo and his Precursors," cited above.

Index

128   INDEX